CAMBRIDGE COUNTY GEOGRAPHIES

SCOTLAND

General Editor : W. Murison, M.A.

# ROSS AND CROMARTY

*Cambridge County Geographies*

# ROSS AND CROMARTY

BY

WILLIAM J. WATSON, M.A., LL.D.

Professor of Celtic Languages, Literature and Archaeology
in the University of Edinburgh

With Maps, Diagrams, and Illustrations

CAMBRIDGE
AT THE UNIVERSITY PRESS
1924

CAMBRIDGE UNIVERSITY PRESS
Cambridge, New York, Melbourne, Madrid, Cape Town,
Singapore, São Paulo, Delhi, Mexico City

Cambridge University Press
The Edinburgh Building, Cambridge CB2 8RU, UK

Published in the United States of America by Cambridge University Press, New York

www.cambridge.org
Information on this title: www.cambridge.org/9781107685345

First published 1924
First paperback edition 2013

*A catalogue record for this publication is available from the British Library*

ISBN 978-1-107-68534-5 Paperback

# PREFATORY NOTE

THE writer is specially indebted to Mr D. Macdonald, Architect, Dingwall, for the plans of Lemlair Church and Fairburn Tower, and to Dr John Horne, LL.D., F.R.S., for invaluable help with the chapter on Geology.

# CONTENTS

|  |  | PAGE |
|---|---|---|
| 1. | County and Shire. Origin of Ross and Cromarty . | 1 |
| 2. | General Characteristics . . . . . . | 4 |
| 3. | Size. Shape. Boundaries . . . . . . | 7 |
| 4. | Surface and General Features . . . . . | 8 |
| 5. | Watershed. Rivers. Lakes . . . . . | 12 |
| 6. | Geology . . . . . . . . . | 19 |
| 7. | Natural History . . . . . . . | 31 |
| 8. | Along the Coast . . . . . . . | 39 |
| 9. | Climate and Rainfall . . . . . . | 48 |
| 10. | People—Race, Dialect, Population . . . . | 53 |
| 11. | Agriculture . . . . . . . . | 56 |
| 12. | Manufactures, Mines and other Industries . . . | 60 |
| 13. | Fisheries . . . . . . . . . | 63 |
| 14. | History . . . . . . . . . | 66 |
| 15. | Antiquities . . . . . . . . | 79 |
| 16. | Architecture—(a) Ecclesiastical . . . . | 88 |
| 17. | Architecture—(b) Castellated and Domestic . . | 98 |
| 18. | Communications—Past and Present . . . . | 109 |
| 19. | Administration and Divisions . . . . . | 112 |
| 20. | Roll of Honour . . . . . . . . | 116 |
| 21. | The Chief Towns and Villages of Ross-shire . . | 128 |

# ILLUSTRATIONS

|  | PAGE |
|---|---|
| Cromarty | 3 |
| Loch Coulin with Ben Eighe in background | 5 |
| A' Bheinn Bhàn | 10 |
| Slioch | 11 |
| At the Black Rock | 14 |
| Falls of Glomach | 16 |
| A peep of Loch Maree from Glen Docharty | 18 |
| Loch Sìonasgaig from Stack Polly | 20 |
| Stack Polly | 23 |
| Liathach | 24 |
| Pterichthys Milleri | 26 |
| Achnasheen Lake Terraces | 28 |
| The Corrie of a Hundred Hills | 30 |
| Wildcat | 33 |
| Oyster-catcher | 35 |
| Castle Leod | 37 |
| Castle Craig | 41 |
| Gairloch-head (c. 1840) | 44 |
| Loch Toll an Lochain | 45 |
| Swordale Beach | 47 |
| Dingwall | 55 |

|  | PAGE |
|---|---|
| Brahan Castle . | 57 |
| Handloom Weaving | 61 |
| Stornoway Harbour . | 64 |
| In the Fishertown, Cromarty | 65 |
| Cadboll Stone . | 67 |
| Shandwick Stone | 69 |
| Urquhart Coat of Arms . | 74 |
| Broch of Carloway . | 77 |
| Penannular Ornament | 80 |
| Eagle Stone, Strathpeffer . | 81 |
| Callernish | 82 |
| Crannog, Kinellan Loch . | 86 |
| Both, Cnoc Dubh, Ceann Thulabhig, Uig, Lewis | 87 |
| Both, Làrach Tigh Dhubhastail, Ceann Resort, Uig, Lewis | 87 |
| Tain from East | 90 |
| Ground Plan of Lemlair Church and of Fairburn Tower . | 93 |
| Fortrose Cathedral . | 96 |
| Ellandonan Castle . | 101 |
| Fairburn Tower | 104 |
| Kilcoy Castle . | 106 |
| Town Hall, Dingwall | 108 |
| Invergordon Ferry . | 110 |
| Hugh Miller . | 120 |
| Hugh Andrew Johnstone Munro | 122 |
| Rev. Gustavus Aird, D.D. | 126 |
| Rev. James Calder Macphail, D.D. . | 127 |
| Sculptured Stone, Dingwall | 129 |

PAGE

Strathpeffer with Ben Wyvis in background    .    .    . 132

Ullapool .    .    .    .    .    .    .    .    .    . 134

Diagrams    .    .    .    .    .    .    .    .    . 136

### MAPS

Rainfall Map of Scotland    .    .    .    .    .    . 51

The illustrations on pp. 3, 5, 11, 14, 16, 18, 64, 65, 90, 96, 101, 134, are from photographs by Valentine & Sons, Ltd., those on pp. 4, 19, 23 are reproduced by permission of Dr W. Inglis Clark and the Clarendon Press, Oxford; those on pp. 10, 24, 28, 30 from photographs supplied by H.M. Geological Survey, Scotland; that on p. 33 is from Dr Jas. Ritchie's *Animal Life in Scotland*; that on p. 35 from a photograph by Mr H. W. Richmond; those on pp. 37, 41, 57, 74, 81, 104, 106, 108 from photographs by Messrs. Munro & Son, Dingwall; that on p. 45 from a photograph by the Rev. A. E. Robertson; those on pp. 47, 61, 77 from photographs by Mr R. Steven, Stornoway; those on pp. 55, 132 from photographs by Mr F. W. Urquhart; those on pp. 67, 110 from photographs by Mr J. Macpherson; those on pp. 80, 82, 86, 87, 129 are reproduced by permission of the Society of Antiquaries of Scotland; that on p. 127 was supplied by Messrs Thomas Nelson & Sons.

# 1. County and Shire. Origin of Ross and Cromarty.

Under the old Celtic system, which lasted on into the twelfth century, Scotland north of the Forth and east of the watershed was divided into provinces governed by nobles of the ancient ruling race, who, though they were nominally under the King of Scotland, were in practice petty kings in their own right. These large provinces were sub-divided into districts under chiefs who were responsible to the ruler of the province. A provincial ruler was styled Mormaer; the chief of a district was styled Tòiseach. This arrangement represented the old tribal organisation.

In the twelfth century great changes and re-arrangements were made. Mormaers became earls (in Latin *comes*); the Toiseach became a thane. The change in style denoted a change in status, for now earl and thane were made to hold title and land from the king, and they were understood to represent the king's authority. In addition to these nobles, other officials were appointed, called *vice-comites*, "deputes of the Earl," in English *sheriffs*. These in theory, though not always in practice, were the servants and agents of the king, whose duty it was to attend to the king's interests, and to see that the king's justice was done, either by themselves or in their presence. The sphere of the sheriff's jurisdiction did not necessarily coincide with that of the earl, for, as one historian puts it, "the realm of the earl or thane was bounded by his own feudal rights of property; the region of the sheriff had a fixed arbitrary limit, the boundaries of the shire or county."

The word "county" comes through Norman French from the Latin *comitatus*, the territory of a *comes*, "count" or earl. Some of the Scottish counties bear the title of an old earldom, for instance Caithness, Ross, Sutherland, Moray, and are therefore properly entitled to be called "Counties." Most of them, however, take their names from the seat of the sheriff, such as Inverness, Nairn, Banff, and are therefore strictly speaking sheriffdoms (*vice-comitatus*). The term *shire* itself is Anglo-Saxon *scir*, *scire*, a charge, a care. It has been borrowed into Gaelic as *sgìre*, a parish, the district allotted to the care of one man for religious purposes.

In 1263 the Earl of Buchan was sheriff of Dingwall. In 1292 William, Earl of Ross, got his lands of "Skey, Lodoux (Lewis), and North Argyll" erected into the sheriffdom of Skye. North Argyll included the west coast of Ross-shire. Later the sheriffdom of Inverness included Ross, Caithness, and Sutherland, as well as Inverness-shire, but in 1503 an Act of Parliament was passed which made Ross a separate sheriffdom, with the sheriff's seat at Dingwall. In 1661 the bounds of Ross were fixed practically as they are now, and Lewis was definitely included in the sheriffdom. In the time of the Earls of Ross, the earl was usually sheriff of Inverness.

The sheriffdom of Cromarty appears to have been instituted in connection with the royal castle there. The first sheriff on record is William de Monte Alto, "vice-comes de Crumbauchtyn" in 1264. The bounds of the

Cromarty

sheriffdom were small, for they did not much exceed
those of the present parish of Cromarty. The first Earl of
Cromarty, Sir George Mackenzie of Tarbat, who was
made earl in 1703, obtained the privilege of having his
various estates, large and small, throughout Ross erected
into a new county of Cromarty, consisting of fourteen
detached pieces. This inconvenient arrangement was
ended in 1891 by the Boundary Commissioners.

By a similar arrangement made in 1476, the Thane of
Cawdor had all his lands in Moray, Nairn, and Ross
made into one thanage of Cawdor, in consequence of
which the parish of Urquhart, in which most of the
thane's Ross-shire lands lay, was reckoned part of Nairn-
shire till 1891. Hence it is that Urquhart is also called
Ferin-tosh, "thane's land," or in Gaelic "an Tòis-
igheachd," the thanedom.

Ross means "promontory," with reference originally
to the large promontory between the Cromarty and
Dornoch Firths, the upland part of which is called *Ard-
rois*, "Height of Ross." Cromarty is in Gaelic *Cromba'*
(open *a*), a difficult name to explain, but certainly containing *crom*, bent[1].

## 2. General Characteristics.

The character and scenery of the mainland part of the
county are exceedingly diversified. Its long double frontage,
to the Atlantic and to the North Sea, gives it two climates,
on the west moist and mild, on the east bracing and dry,
and in the uplands cold. The west coast, much and

---

[1] See *Place Names of Ross and Cromarty*.

Loch Coulin with Ben Eighe in background

deeply indented by sea-lochs and close to the mountains, offers a striking contrast to the fertile plains and sunny slopes and valleys of the east. Only about seven per cent. of the whole surface is arable. The mountainous interior rises to peaks of nearly 4000 feet, with endless variety of glens, lakes and streams in between. The upland pastures at one time fed great herds of the shaggy wide-horned Highland cattle. Towards the end of the eighteenth century these began to be displaced by sheep, which in their turn have made way for preserves of deer and grouse. Easter Ross, the Black Isle, and the lower parts of Mid Ross are notable farming districts. Smaller farms and crofts are most numerous in the higher parts and on the west coast. In certain places land once tilled has been annexed to deer-forests. On the other hand, during last century much land was reclaimed and improved, especially by Sir Alexander Matheson in Ardross. Many districts are still remote from railways; Ullapool, for example, is thirty-two miles from the nearest station; and it is rather remarkable that there is still no railway between Dingwall and Cromarty. The deficiency of communication on the west coast is only partly relieved by a rather unsatisfactory steamer service.

On the east side, the way by land between north and south has always of necessity been round the heads of the firths of Beauly and Cromarty, while in addition Dingwall commands the route between east and west. The Beauly river formed a natural boundary which marks the southern limit of Norse influence. It is not surprising, therefore, to find important ancient and mediaeval strongholds in this region. The pre-historic forts on the Ord of

Kessock and on Knockfarrel were the forerunners of the feudal strengths of Redcastle and Dingwall. Similarly the entrance to the Cromarty Firth was guarded by the castle of Cromarty and the earlier Dunskaith on the Nigg side. The experience of the late war has again shown on a larger scale the vital importance of Cromarty and its firth. On the west, the castles of Strome and of Ellandonan were of first-rate importance as commanding the narrow seas and the passes to the east.

## 3. Size, Shape and Boundaries.

The total area of the county, including islands, is 3260 sq. miles, of which 103 are water. The greatest length, from north to south, is about 67 miles and the greatest breadth is about 75 miles. The shape of the mainland part has been compared to a fan, of which the part between the Moray Firth and the Dornoch Firth forms the handle, while the seven or eight promontories on the west, divided by sea-lochs, represent the ribs extended. On the north, the boundary between Ross and Sutherland is formed chiefly by the river Oykel and its estuary, the Kyle of Sutherland. Beyond the source of the Oykel, the boundary runs in an irregular curve north-westward by Loch Veyatie, Loch Fionn and the river Kirkaig, meeting the sea at Inverkirkaig. On the north-east there is the Dornoch Firth, and on the east the North Sea. The south-east side from Tarbat Ness is bounded by the Moray Firth to Chanonry Point, thence by the Inverness Firth to Kessock Ferry, and thence by the Beauly Firth to Tarradale. Thereafter the march

between Ross and Inverness-shire runs irregularly west by
south to the head of Loch Monar. Then it goes south-
west to A' Bheinn Fhada (Ben Attow), where it wheels eastwards to include Glenshiel, and so westwards along the
watershed between Glenshiel and Glenelg till it reaches
the sea near the north end of Kyle Rhea. The whole of
the west side faces the Minch till the march of Sutherland
is reached at Inverkirkaig.

Lewis, with the Atlantic on the west and the North
Minch on the east, contains 683 sq. miles. Its greatest
length is about 43 miles, and its greatest breadth 28 miles.
It is separated from Harris, the southern part of the island,
belonging to Inverness-shire, by Loch Resort on the west
and Loch Seaforth on the east, and the isthmus, 6½ miles
broad, between their heads. Lewis is shaped like a kite,
with the Butt as apex. On the west side it is deeply
indented by Loch Roag.

## 4.  Surface and General Features.

The plain of Easter Ross slopes northwards towards a
low smooth ridge of hills running from Tain towards
Ardross, and forming as it were the backbone of the
promontory which gives its name to the county. The
Black Isle, which is really a peninsula, rises on both sides
to the ridge known as Ard Mheadhonach (mid-height),
the highest point of which is 838 feet. Its western end
gives a magnificent view of the really Highland part of
the county, from Wyvis (3429) in the foreground to
Sgùrr na Lapaich (3773) on the left. Wyvis itself is seen
rather as a massive range than a single mountain, domi-

nating the landscape and sheltering by its bulk the low
ground to the east. Its most notable neighbours are Carn
Chuinneag (2749) with its shapely double peak; Sgùrr
Mór in Fannich (3637), the highest point north of Inverness, from which on a very clear day one can see most of
Scotland north of the Grampians ; and Sgùrr a' Mhuilinn
(2845) at the head of Strathconon. Along the watershed
from south to north are Sgùrr nan Conbhairean (3634),
A' Bheinn Fhada (3383), Màm Sabhal (3862) with its
neighbour Carn Éite (3877), Maoil Lunndaidh (3294),
Mórusg (3026), Fionnbheinn (3060), and Beinn Dearg
(3547). West of the watershed there are the stately
"Five Sisters" at the head of Loch Duich, namely, Sgùrr
na Mormhoich (2870), Sgùrr nan Saighead (2750), Sgùrr
Udhran (3505), Sgùrr nan Carnach, and Sgùrr nan
Cisteachan Dubha (3370). Maoil Cheanndearg (3060)
and Sgùrr Ruadh (3141) in Lochcarron are of the ruddy
Torridonian rock, while their neighbours Na Cinn Liath
or Grayheads (3034) are quartzite on top. Liathach
(3358), Spidein a' Choire Léith (3456), Beinn Eighe
(3217) and Ruadh Stac (3309) are all near the head of
Loch Torridon. The finest hill in Applecross is A' Bheinn
Bhàn (2936). Sleaghach (Slioch) near the head of Loch
Maree is a great truncated cone (3217), deeply fluted
with water-worn gullies. Mullach Coire mhic Fhearchair
(3250), Sgùrr Bàn (3194), Badhaisbheinn (2869) and Beinn
Ailiginn (3232) are in Gairloch. An Teallach (3483), at
the head of Little Loch Broom, is a wild serrated range
resembling the Coolins of Skye. In Coigach are A' Bheinn
Mhór (2438), Cuthail Bheag (2523) and Cuthail Mhór
(2786) ; the two latter are fine shapely rounded hills.

A' Bheinn Bhàn

For so mountainous a county, the average height of the valley-floors is surprisingly low. Achnasheen, on the watershed between east and west, is only about 500 feet above sea-level, Strathconon reaches 400 feet only at a distance from the sea of 18 miles. Strathcarron (Ardgay) is 300 feet at its head, 10 miles up; 8 miles higher, Srath Cuilionnach is 480 feet. Few of the west coast glens

Slioch

reach 400 feet at their heads; most of them fall far below that elevation. No valley in Ross-shire compares in height with, for example, Glen Lyon with its elevation of 500 to 1100 feet, or Strathdearn with 1000 to 1300 feet.

The surface of Lewis is rather featureless, though its brown heathery moors are not without charm. The only hilly part is the parish of Uig, in the west, with Mealasbhal

(1885), Taithabhal (1688), Cracabhal (1682) and others rather lower. In the northern part Mùirneag (1807) forms a very conspicuous mark. The southern part contains innumerable fresh-water lochs, mostly small. Most of the Lewis hill names end in *-bhal*, *-val*, from Old Norse *fjall*, a hill, while the larger loch names end in *-bhat*, *-vat*, Old Norse *vatn*, water.

## 5. Watershed. Rivers. Lakes.

The main watershed runs north and south, being a section of the great divide between the east and the west, known of old as Drum Alban, the dorsal ridge of Scotland. Beginning at Ben More of Assynt in the north, it runs southwards to Loch Droma, near Braemore, at the highest point of the road from Garve to Ullapool. Thence it goes by Loch Fannich, and south-west to the head of Loch Chroisg. Thereafter it goes south by Mórusg to the head of Glen Fhiodhaig, and continues in the same direction by the head waters of Loch Monar and Glen Siaghaidh to A' Bheinn Fhada (Ben Attow), striking the Inverness march at Maoil Cheanndearg (3342) near the head of Glen Cuaich. The general direction of this great watershed is south by west. A subsidiary watershed between the Dornoch Firth and the Cromarty Firth runs from the head of Strathrusdale eastwards to Tain, a distance of about 18 miles.

Most streams that rise to the east of the principal watershed find their way either to the Dornoch Firth or to the Cromarty Firth. The chief rivers that fall into the former are Oykel and Carron. The Oykel, from Loch

Ailsh, receives the Eunag at Amat, near Oykel Bridge, and enters the sea at Bonar Bridge, but is tidal for about 10 miles above that point, and during this part of its course is called the Kyle of Sutherland. The Carron rises in Glen Beag near the border of Loch Broom and after a course of about 20 miles joins the sea at Bonar Bridge. Its principal tributaries are the Blackwater from the left and the water of Glencalvie from the right, both of which join it near Amat (Norse *á-mót*, water-meet). The most easterly of the streams that flow into the Cromarty Firth is the Balnagown Water, from Srath-uaraidh (Strathrory), "landslippy strath," so called with reference to the steep clayey banks of the stream. At the mouth of the strath, the river has cut its way through a deep narrow ravine in the sandstone. Further west the Averon enters the firth near Alness. High up it passes through Loch Moire, three miles below which it receives the Blackwater from Strathrusdale, and soon after plunges into a deep valley with Ardross Castle on its left. The river Glass flows through Loch Glass and Glenglass into the firth near Evanton. It is remarkable for the tremendous chasm a mile long and about 130 feet deep which it has cut out of the conglomerate rock between Glenglass and Evanton. This chasm, known as An t-Allt Grànda, "the ugly precipice," is in places only about 12 feet wide, and its sides are perpendicular. Further up the river are the beautiful falls of Coneas. A little to the west is the Skiach Water from the skirts of Wyvis. The small Peffery from Strathpeffer gives Dingwall its ancient name of Inbhir Pheofharain, "Peffer Mouth," as it is always called in Gaelic. At the head of the firth enters the river Conon, the largest of the Ross-shire rivers. In early times, according to geologists, the Conon flowed through

**At the Black Rock**
(*on the River Glass*)

the Souters into the Moray Firth, so that the Cromarty Firth may be regarded as primarily its estuary. The Conon might naturally be expected to be the river of

Strathconon, but now at any rate the name applies to the
stream which, rising from Loch Chroisg on the watershed
near Achnasheen, flows through Strathbran, and is usually
called the Bran till it enters Loch Luichart.  Thereafter
to the sea it is called the Conon.  Its chief right bank
tributary is the Meig, rising at the head of Glen Fhiodhaig and flowing through Loch Beanncharan and Strathconon to Scatwell.  At Urray the Conon is joined by the
Orrin, a rapid stream which rises a little to the north of
Loch Monar and flows through Loch na Cuinge and
another small loch, and then through Glen Orrin, being
joined below Falls of Orrin by the water of Glen Gowrie.
On its left side near Moy Bridge the Conon receives the
Blackwater, which rises on the borders of Loch Broom
as the Glascarnoch river, and then flows through Strath
Garve and Loch Garve, forming the Falls of Rogie near
Strathpeffer.  On the south side of the Black Isle, the
Eathie Burn, near Cromarty, is noted for its fossil beds,
while the Rosemarkie Burn (the Marknie) has immense
cliffs of boulder clay.  Both these fall into the Moray
Firth.

As the main watershed is nearer the western sea, the
streams on the west are shorter and smaller, and they
drain comparatively small areas.  Loch Duich receives the
rivers Shiel, Lichd and Conag, while the Ling and the
Elchaig fall into Loch Long.  On the Glómach, a tributary
of the Elchaig, are the lofty Falls of Glómach.

The Carron rises near Achnasheen, flows through
Loch a' Ghobhainn, Loch Sgamhain and Loch Dughaill,
and through Glen Carron and Strath Carron into Loch
Carron.  From Loch Maree (formerly Loch Ewe) the

short river Ewe enters Poolewe.  Further north, the
Little Gruinard river, the Inver-riavenie river and the

Falls of Glomach

Gruinard fall into Gruinard Bay, while the Strathbeg
river falls into Little Loch Broom.  Abhainn Bhraoin

(The Broom) rises in Lochaidh Bhraoin and flows through Glen Braoin into Loch Bhraoin (Loch Broom). In Coigach the river Pollaidh (Polly) drains Loch Sìonasgaig into Enard Bay. The Kirkaig comes from Loch Mhiadaidh (Veyatie) and enters the sea at Inverkirkaig.

The largest and the most beautiful of the lochs is Loch Maree, 12 miles long, with a number of islands. It was formerly called Loch Ewe and came to be called Loch Maree from the island Eilean Ma-ruibhe, which contained a chapel named after Maol-rubha or Maol-ruibhe, the founder of the monastery of Applecross. Similarly Loch Moire, in Alness parish, is named after the Virgin Mary, whose ancient chapel stands at its head. With regard to Loch Glass, Mr James Fraser, the minister of Alness writes about the year 1730, "the river (Glass) is not sonsy, nor yet the loch from which it comes, being Loch Glaish 3 miles in lenth. But they think the water is sanctifyed by bringing water to it from Lochmoire." Into Loch Ussie, near Dingwall, Coinneach Odhar, the noted wizard, threw his magic stone of divination. Loch Kinellan, near Strathpeffer, contains a crannog or lake-dwelling, recently investigated. Similar structures are found in Loch Beanncharan, Loch Tollie in Gairloch, and in many other lochs. Of the larger lakes Loch Fannich, 6 miles long, and 821 feet above the sea, comes next in elevation to the Lochan Fada in Gairloch which is 1000 feet, but some lochlets are still higher. The district of Coigach is rich in lochs, such as Loch Sìonasgaig (on some maps L. Skinaskink), Loch Osgaig (on maps L. Owskeich), Loch Bad a' Ghoill and Loch Lurgain. There are also Loch Achall, near Ullapool;

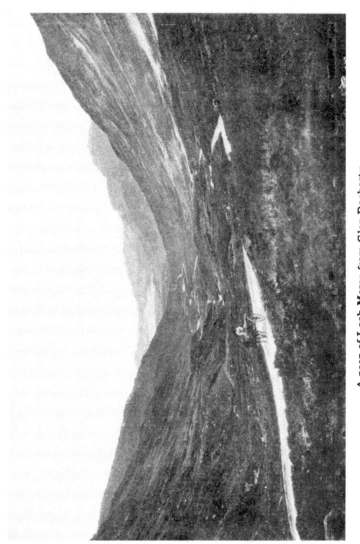

A peep of Loch Maree from Glen Docharty

Loch Luichart, Loch Garve, Loch Chroisg, Loch Clair, Loch Coulin, Loch na Sealga, Fionnloch, and a host of others.

The watershed of Lewis runs from north-east to south-west, keeping about the middle of the island, except in the north, where it approaches the eastern side. Small streams are numerous, but there is no large river. The Laxa (salmon-water) and the Greeta or Creed are near Stornoway. Of the others, the Grimersta river, which enters the head of Loch Roag, is noted for salmon. The largest loch is Loch Langavat (Longwater), about 6 miles in length; another loch of the same name in the northern end is less than a mile long. The parish of Lochs (Sgir nan Loch) is so called from the almost countless number of its lochlets.

# 6. Geology.

Geology deals with the history of the earth's crust. It describes the cycle of processes by which rocks are formed, how their mineral constituents are worn away and re-deposited, thus giving rise to younger strata. It explains how the features of our hills and valleys have been evolved by the agents of denudation. From a study of the relative position of the rocks and the fossils found in some of the beds, geologists have been able to determine the order of their succession in time. The fossils furnish valuable evidence of the life that existed on the earth at various stages of its history, and give a clue to the order in which the species of plants and animals

Loch Sionasgaig from Stack Polly

appeared, how they were modified, and, in many cases, disappeared.

Rocks are of three kinds, (1) aqueous or sedimentary, (2) igneous, (3) metamorphic. The sedimentary types have certain characteristics whereby they can be readily recognised in the field. They consist of conglomerates, sandstones, and shales, composed respectively of gravel, sand, and mud, derived from the disintegration of older rocks, and laid down under water. In some formations they are associated with limestones, containing organic remains, and seams of coal made up of petrified vegetable matter. The loose sediments were converted into hard rocks partly by pressure and partly by cementing material such as carbonate of lime and silica. The members of this division are stratified, the lowest layers when the sequence is normal being the oldest, but when the beds have been overfolded by movements of the earth's crust, the original order of succession is disturbed.

The igneous rocks are due to intense heat in the interior of the earth. The molten material may be ejected at the surface from volcanoes, thereby cooling quickly in the form of lava, or it may consolidate slowly beneath the surface with coarse crystallisation as granite.

The metamorphic rocks include types of sedimentary and igneous origin which have undergone change, caused partly by heat and partly by pressure. By these processes sandstone has been converted into quartzite, limestone into marble, shale into slate or mica-schist, and granite into gneiss. This division comprises the great development of crystalline schists and gneisses which form the largest portion of the counties of Ross and Cromarty.

The following table gives the order of succession of British stratified rocks :

Recent and Post-Glacial (Neolithic).
Pleistocene or Glacial (Palaeolithic).

Tertiary or Cainozoic
- Pliocene
- Miocene
- Eocene

Secondary or Mesozoic
- Cretaceous
- Jurassic (Lias and Oolites)
- Triassic

Primary or Palaeozoic
- Permian
- Carboniferous
- Old Red Sandstone and Devonian
- Silurian Upper and Lower
- Cambrian

Pre-Cambrian
- Torridonian
- Archaean

The Lewisian Gneiss forms the Archaean floor in the North-West Highlands. The rocks consist mainly of an assemblage of gneisses of igneous origin with a limited development of schists, representing altered sedimentary rocks, including quartz-schists, mica-schists, graphite-schists and limestones. These rocks are traversed by intrusive dykes and sills of various igneous materials. The members of this division occupy nearly the whole of the Outer Hebrides ; on the mainland they appear at Letterewe and Poolewe, extending southwards to Gairloch and at intervals to Loch Torridon. The gneisses with their intrusive dykes and the schists of sedimentary origin have been thrown into sharp folds, which are well seen north and south of Loch Maree.

The Archaean floor, worn into hills and valleys by prolonged denudation of the old land surface, is covered unconformably by a great pile of red sandstones and grits, several thousand feet in thickness, which form the Torridonian system. These materials were accumulated under continental conditions, the prevalent felspar (microcline) in the grits being perfectly fresh. Some of the pre-Torri-

Stack Polly

donian valleys are filled with these sediments to a depth of about two thousand feet. Remnants of this extensive formation still survive as isolated hills like Stack Polly, or as mountain groups like the hills of Applecross or Coigach. The terraced arrangement of the sediments is well displayed in Liathach at the head of Loch Torridon and on Slioch at the head of Loch Maree.

Next in order come the Cambrian rocks which form a narrow strip of country extending from Loch Eriboll in Sutherland, thence across Ross-shire by Ullapool and Kinlochewe to Loch Kishorn. The prominent members of this system are quartzites, dolomites and limestones, which are proved to be of marine origin by the occurrence in certain bands of marine fossils. Well-preserved trilobites

Liathach

have been found in shales on Meall a' Ghiuthais—a hill near Kinlochewe at the head of Loch Maree. The quartzites are conspicuous features in the landscape ; they cap the mountain tops and give rise to long dip slopes towards the east. Within this narrow belt the rocks have been folded, piled up, and driven westwards in successive slices by earth-movements in post-Cambrian time. The great displacements of Lewisian Gneiss, Torridon Sandstone and Cambrian strata are clearly displayed between Kinlochewe and Loch Kishorn.

To the east of the Cambrian strip lies a belt of mountainous country—about thirty miles wide—stretching from Kinlochewe to beyond Garve, which is occupied by crystalline schists belonging to the Moine Series of the Geological Survey. They comprise quartzose granulites, mica-schists, and garnetiferous muscovite-biotite gneiss, representing what were originally arenaceous and argillaceous sediments. They are pierced by intrusions of older foliated granite and epidiorite, and younger granite masses introduced after the foliation of the schists. The foliated granite is largely developed round Carn Chuinneag and at Inchbae west of Ben Wyvis ; the later unfoliated type, between Strathrusdale and the Dornoch Firth. Along some of the sharp folds in the Moine Series, gneisses and schists occur, that resemble certain types of Lewisian Gneiss in the western seaboard of Sutherland and Ross.

The belt of richly cultivated ground in Easter Ross is formed of Old Red Sandstone, whose component strata rest unconformably on a very uneven surface of the crystalline schists. The western boundary extends from a point near Beauly north by Castle Leod to Edderton near the Dornoch Firth. Outlying patches of this formation surrounded by crystalline schists occur in Strath Rannoch north-west of Ben Wyvis, where the basal conglomerates contain numerous blocks of foliated granite derived from the masses at Carn Chuinneag and Inchbae. An interesting feature of the higher conglomerates at Alness and Evanton in the main belt is the occurrence in them of fragments of Cambrian limestone and Torridon sandstone which have travelled far from their parent sources.

The Old Red Sandstone strata in Easter Ross form a syncline or basin, the lowest beds being exposed on the western rim north of Beauly and on the eastern rim in the Sutors of Cromarty. The highest beds, consisting of red and grey sandstones, occur along the Millbuie ridge

Pterichthys Milleri
(*A fossil fish of the Old Red Sandstone*)

in the Black Isle. Two prominent members of this succession of strata have become widely known ; the fetid bituminous shales from which spring the mineral waters of Strathpeffer, and the famous fishbeds at Cromarty. The latter locality, made classic by the researches of Hugh Miller, has yielded a rich suite of fish remains (ichthyolites) which are now referred to the Middle Old Red Sandstone.

The Secondary rocks, though originally widespread, are now confined to small areas on the east and west borders of the county. They occur on the east coast at Eathie Bay south of Cromarty, at Port an Righ and Cadha an Righ north of the Sutors, where they consist of sandstones, clays, and limestones, yielding fossils ranging in age from the Lower to the Upper Oolite. At Cadha an Righ a thin coal seam appears in the Lower Oolite on the same horizon as the workable coal seam at Brora in Sutherland. At these localities the Secondary rocks have been let down to the south-east against the crystal-line schists and Old Red Sandstone by the great fracture that traverses the Great Glen and skirts the coast line from Kessock to Tarbat Ness.

On the west coast an important area of Liassic strata, consisting of sandstones, grits, conglomerates with thin coal seams, and limestones, appears at Applecross where they have been let down among the mountains of Torridon Sandstone by powerful faults or fractures. Farther north a narrow strip of Secondary rocks, composed of Triassic and Liassic sediments, stretches from Sand in Gruinard Bay to Aultbea. Like the beds at Applecross they are bounded by two great faults which have thrown them down against the harder Torridonian strata, whereby they have been protected from denudation.

Occasional records of the great outburst of volcanic action in Tertiary time in the West Highlands are to be found in Applecross where the Torridon Sandstone and Liassic strata are traversed by sheets and dykes of dolerite and basalt belonging to this period. Near Toscaig Bay south of Applecross two small volcanic vents pierce the

Achnasheen Lake Terraces

Torridon Sandstone which are filled with agglomerate containing blocks of sandstone and basaltic lapilli.

The evidence relating to the Glacial period, furnished by the glaciated rock slopes, ice-markings, boulder clay, moraines, and carried boulders, is of special interest. During the great glaciation the whole region was enveloped in ice which moved west and north-west to the Sound of Raasay and the Minch, and eastwards towards the Moray Firth. At this stage the ice-shed lay several miles to the east of the existing watershed at Loch Droma. At a later stage the Fannich mountains and the mass of high ground near Beinn Dearg became centres of dispersion and the tops of the western mountains rose above the ice-field. As the cold became less severe each independent area of high ground radiated glaciers which followed the trend of the valleys.

Boulders of the Inchbae foliated granite have been carried across the main watershed of the county, north-west to Loch Broom and eastwards across the Black Isle and the Moray Firth to the plain of Moray and onwards to Banffshire. Blocks of crystalline schist from Central Ross have been borne to a height of nearly 3000 feet on the western mountains where they rest on undisturbed Torridonian and Cambrian strata. Some of the blocks have ascended about 1500 feet above their parent source. Moraine mounds and ridges strewn with boulders form prominent features in the landscape, as between Kinlochewe and Loch Torridon where lies " the corrie of a hundred hills " (Coire Ceud Chnoc).

Well-marked terraces indicating former levels of the sea occur at intervals on the east coast between Dornoch

The Corrie of a Hundred Hills

and Beauly Firths, and on the west coast in Applecross, in Loch Carron and Loch Kishorn. The lower raised beaches range in height from eleven to thirty feet and the higher ones from fifty to one hundred feet. The fertility of the lowland belt in Easter Ross is due to the materials accumulated there during the Glacial period and to the deposits of the various raised beaches.

## 7. Natural History.

The British Isles were not always islands, nor were they always separated from each other. At a period not so very distant, as geologists reckon time, Ireland was joined to Great Britain, and both were connected by land with the continent of Europe. This was the position at the close of the great Ice Age and for a long period thereafter, during which what is now the bed of the North Sea was a great rolling plain, traversed from south to north by a mighty river, formed by the combined waters of the Thames, the Rhine, and other affluents. Throughout this period animals and plants spread without hindrance from the continent into Britain. First came the arctic fauna and flora; later, as the climate grew milder, came plants and animals adapted to a temperate climate. As the land gradually sank, Ireland became separated from Britain before Britain was separated from the continent: the Irish Sea existed before the English Channel. Thus immigration by land into Britain continued longer than into Ireland, and this is the reason why Ireland has fewer mammals and reptiles than Britain, while Britain again has fewer than the continent. As the

Minch is about the same depth as the Irish Sea, it is probable that Lewis became an island about the same time as Ireland, and this may be a reason why Lewis has fewer plants and animals than the mainland.

Among the animals of the early period were the elk and the reindeer. Hugh Miller records the finding of a large fragment of elk's horn in a mossy ravine near Cromarty; another was found in Strath Halladale in Sutherland. Remains of the reindeer have been found in the Mormhoich Mhor near Tain, and also in certain Caithness brochs. According to the Orkneyinga Saga, the Earls of Orkney used to hunt the roedeer and the reindeer in Caithness. Wolves were by no means scarce in Scotland in the seventeenth century. Of Strathnaver, for instance, it is reported, "especially heir never lacks wolves, more than ar expedient." When the wolf became extinct in Ross is not known, but tradition of them lives still. At Coulin, for instance, they used to cross the river by a ford still called Pait nam Madadh, the Wolves' Ford. In the middle of last century, the badger (broc), marten (taghan), and polecat (feòcullan) were common in the west, and they are not yet extinct.

The true wildcat is still found in several parts of the county, but many supposed wildcats are either the domestic cat run wild or a breed between it and the true wildcat. Foxes are fairly common in the wilder parts. Otters haunt many of the rivers, and are also found along the rocky coasts. Weasels and stoats are found all over, but the former are much the more numerous. Squirrels, which cause so much damage to fir trees, seem to be less plentiful than formerly. The brown rat is too common everywhere; the more harmless black rat is rare. Of the mouse tribe we have the common house-mouse, the long-tailed field-mouse, which haunts gardens and eats bulbs and fruit, the short-tailed field-mouse found in cornfields, and the little shrew-mouse, which lives on worms. Moles are found wherever earth

Wildcat

is. The rabbit is an importation—in Gairloch they were introduced about 1850—and always tends to become a pest, destructive of grass, corn, and turnips. Under the game laws the rabbit is vermin. Brown hares are fairly common on the low ground, and there are large numbers of mountain hares, blue in summer, white in winter. The sacred red deer is monarch of about 900,000 acres of the mainland and Lewis. One often hears the plash of the water-rat or vole and may catch a glimpse of his trail in the water, but he is too shy to be often seen.

Of the 368 species of birds in Britain, 153 have been noted by Mr J. H. Dixon in Gairloch alone, either as nesting or as visitors. Among the birds of prey, the golden eagle is found in the wilds, and the white-tailed or sea-eagle on the west coast. Of hawks, there are the peregrine falcon, the merlin, the kestrel, the sparrow-hawk, the kite, the buzzard, and the hen-harrier. Ravens are fairly common, and the hoodie crow is often seen. The tawny owl, the short-eared owl, and the white or barn owl are all found; the long-eared owl is seen some-times as a visitor. There is also the night-jar, with his peculiar purring note. Rooks are common, and so are jackdaws, but the magpie is becoming rather rare. On the high hills the ptarmigan and the snow-bunting appear, and on the moors there are the curlew, the snipe, the lapwing or peewit, the golden plover, and the red grouse. Herons may often be flushed from their fishing in the quiet stream-pools. The capercailzie is not found, but there are blackcock and partridges on the foothills and low ground. Among the countless birds that haunt the shores or sea are many kinds of gulls, the kittiwake, skua, razorbill, stormy petrel, arctic tern, dotterel or ringed plover, dunlin (on beaches), the great cormorant and the green cormorant or shag, the gannet or solan goose, puffin, guillemots of various kinds, great northern diver, oyster-catcher, and sea-swallow. Many of these come inland. Of the goose kind, there are the greylag goose, which nests on islands in fresh-water lochs and does great damage to corn, the brent goose (rare), the sheldrake, mallard or wild duck, teal, and goosander. Wild swans may be seen occasionally, as on Loch Maree. The kingfisher, most brilliant of all native birds, is rarely seen. The harsh cry of the corncrake or landrail is often heard in the east, seldom in the west. Songbirds are numerous

Oyster-catcher

in the lower parts, but the lark is less common in the west than it used to be, and is seldom heard in the uplands. The mavis or song-thrush and the blackbird are common, and the greenfinch, or green linnet, the chaffinch, the bullfinch, the goldfinch (occasionally), the siskin, and the starling are found, along with many others. Wood pigeons are numerous, and rock pigeons are plentiful in the rocks and caves of the coast. Wood-cocks are common on the low ground.

The commonest reptile is the adder, whose bite has often been fatal. Lizards are not very common. Frogs, toads, and newts represent the amphibious creatures. In the lochs and streams there are salmon and sea-trout, brown trout, char, pike, and eels. Sea fishes are too numerous to mention in detail.

Remnants of the old pine forests survive in Coulin and perhaps elsewhere, but the once famous forest of Coire Mhàileagain in the heights of Kincardine, part of which grew thick and tall in Hugh Miller's time, has now only withered stumps. This splendid species has ceased to be self-propagating, and it has been suggested as one possible cause of this that the deer nibble the young plants as they appear. The common Scots fir is far inferior to the old pine of Caledonia. The chief other native or indigenous trees are the oak, birch, hazel, aspen, rowan, bird-cherry, hawthorn, blackthorn, alder, willow, and holly, of which the last is not very common. Of trees not indigenous, the larch was introduced about the middle of the eighteenth century, and does well on the eastern side, sowing itself readily. Till recent times a magnificent larch, one of the first planted in Ross, grew at Novar. It had to be felled owing to the danger to adjacent houses, had it been blown down. Spruce thrives well in the woods of Ardross and elsewhere. Beech, ash, chestnut, and lime-trees flourish on the low grounds, especially about Castle Leod, Balnagown, and New Tarbat. The tall Lombardy poplar is a feature of Dingwall. Whins and broom flourish in the east, but only on fairly low ground ; in the west they do not seem to be at home. Juniper has much the same range, but is not so common or so luxuriant. Around Tomatin, Inverness, juniper thrives freely at a height of well over 1000 feet; in Ross it is hardly found above 500 feet. Wild rasps grow and ripen; brambles ripen seldom. On boggy moors the chief shrub is the fragrant bog-myrtle or gale (*roid*); on heathery slopes are the cranberry (*braoileag*),

Castle Leod

red bearberry (*grainnseag*), black heathberry (*fiantag*), blaeberry or bilberry (*dearcag*), and, high up, the cloudberry (*oighreag*).

As might be expected from its extent and physical characteristics, the county has a large variety of flowering plants, including most of the familiar and some of the rarer species. Fortrose, for instance, is said to be the only part of Britain in which is found the white butterwort

(*Pinguicula alpina*), an insectivorous plant. Of the many
species of the lower grounds may be named the saxifrages
(eight varieties), the commonest being the golden saxi-
frage, fringing the burnsides; ranunculi—celandine, water
crowfoot, spearwort, buttercup; cow parsnip (Gaelic,
*giùran*), a tall umbelliferous plant, sometimes mistaken for
the wild hemlock; honeysuckle (*Lonicera Periclymenum*);
ragged robin; the wood anemone (*Anemone nemorosa*); and
tropaeolum, which is claimed as a specialty of Strathpeffer.
In the uplands, wild thyme is common on dry banks, and
several varieties of orchids, purple, white, and spotted,
bloom on the hard pastures. The foxglove (*Digitalis
purpurea*) often grows at considerable heights in favourable
situations. The commonest plants of the moors are the
cotton grass (Gaelic, *canach*), the little fragrant yellow-
headed bog asphodel (*Narthecium ossifragum*), and several
varieties of the insectivorous sundew (*Drosera*). The tarns
often have water-lilies, and trefoil (Gaelic, *lus nan laogh*)
grows in slow running moor channels. Other alpine
plants are alpine meadow rue (*Thalictrum alpinum*), alpine
scurvy grass (*Cochlearia alpina*), and *Salix herbacea*, a dwarf
willow. *Saussurea alpina*, a rare plant with a strong helio-
trope odour, is found on Sgurr Udhran, Loch Duich.

Wester Ross has fewer species of plants than Easter
Ross, and Lewis has fewer still. The pyramidal bugle
(*Ajuga pyramidalis*) is found on the river Creed near
Stornoway, and sea samphire grows on the cliffs at Man-
garsta. Water-lilies are found in some of the lochs; water-
lobelia, bladder-wort, and a species of sundew are common[1].
The royal fern is frequent, and liverworts are abundant.

[1] Mr W. J. Gibson in *A Guide to Stornoway*.

The only woods in Lewis are those in the neighbourhood of Stornoway Castle.

In Lewis the otter, seal, shrew mouse, common mouse and rat are found. There is also a blue-black rat, known as the "Egyptian rat," which has made its appearance within recent years. There are red deer, but no roe-deer. Foxes, wildcats, and pole or marten cats have all been killed out. Hares are found, and rabbits were introduced about fifty years ago. The frog, toad, newt, weasel, and badger are absent, and there are no black game or partridges. Whether adders appear is doubtful, but the blindworm occurs[1].

# 8. Along the Coast.

The Dornoch Firth is geologically the old bed of the Oykel, as the Cromarty Firth is the old bed of the Conon. We start from Bonar Bridge, which connects Ross and Sutherland. Here on the south side is the village of Ardgay, and hence to Tain road and railway run parallel to each other near the shore. The most prominent feature is the bold granite bluff of Struie Hill on the right. The shore is low, edged by a narrow strip of good land which widens to a deep hill-girt embrasure at Edderton. Three miles from Tain a long sandy coulter-shaped spit projects into the firth to within about three-quarters of a mile of

---

[1] See further Harvie-Brown's *Fauna of the Outer Hebrides*. Since its publication some changes have taken place, especially addition of birds owing to the trees in the Castle grounds. For this and the above I am indebted to Mr W. J. Gibson, Rector of the Nicolson Institute, and to Mr Mackenzie, Royal Hotel, Stornoway.

the opposite side; here is the Meikle Ferry, in Gaelic
Port a' Choltair, Coulter Ferry. Off Tain the entrance
to the Dornoch Firth is obstructed by the Gizzen Briggs,
a dangerous bank, whose roar may be heard for many
miles ; in tradition the Briggs are the remains of a land
bridge that once extended almost from Ross to Sutherland.
The great sandy plain of the Mormhoich Mhór and the
Fendom, eastwards of Tain, was once covered by the sea,
and the old beach is very well marked, reaching about
3 miles inland. Now, however, the sea is again gaining
on the low sandy shore. Between this and Tarbat Ness
are the fishing villages of Inver and Portmahomack, while
on the Ness itself is an important lighthouse. Rounding
the point, we find the coast forming almost a straight line
south-west to Kessock Ferry, a distance of about 31 miles,
much of which presents a bold steep or even precipitous
front to the Moray Firth. Passing the fishing villages of
Rockfield, Balintore and Hilton, we reach the entrance
to the Cromarty Firth, guarded by the Sutors of Cromarty,
two great headlands 7 furlongs apart. The precipices
and caves of the South Sutor were the scene of Hugh
Miller's boyish adventures described so vividly in his
*Schools and Schoolmasters.* There are also several rather
extensive caves in the southern face of the North Souter,
eastwards of the opening of the firth. The Cromarty
Firth itself runs about 20 miles to the north-west, and
its eastern part is one of the finest natural harbours in
the world. During the Great War it was of immense
advantage to our fleet, and the quiet towns of Invergordon
and Cromarty became busy and stirring places. On the
north side, Dingwall, the county town, stands at the head

of the firth, with Wyvis in the background.  Further east
are Evanton and the large village of Alness, to the north-
west of which the steep bold hill of Fyrish stands conspicuous.  On the south side, nearly opposite Kiltearn
Church, stands what remains of Castle Craig, once a

Castle Craig

residence of the Bishop of Ross.  Outside the firth the
coast continues straight to Chanonry Point, a long tongue
of land which reaches out nearly opposite to Ardersier
Point, about a mile away.  This, says tradition, is part of
a road nearly, but not quite, finished in one night by

fairies at the bidding of the wizard Michael Scot. Here
Rosemarkie and Fortrose stand close together. The former
was the site of a sixth century church founded by St Mo-
Luag of Lismore. Fortrose was noted for its beautiful
cathedral, part of which still stands. Further along is Avoch,
a thriving fishing village, and a mile further is Ormond
Hill, with the traces of a great feudal castle. Munlochy
Bay, 2 miles long, has Munlochy village at its head,
and on its southern side, near the entrance, Creag a'
Chobha (Craigiehow) with a cave in which the Fiann lie,
and a dropping well which was reputed to cure deafness.
But a Gaelic proverb says, "ged is mór Creag a' Chobha,
is beag a feum," though Craigiehow is big, small is its
use. At Kessock Ferry, 3 furlongs wide, is the village of
Craigton, occupied mainly by pilots of the Moray Firth.
On the Ord of Kessock there are remains of a large
vitrified fort. Entering the Beauly Firth we observe an
islet, Carn Dubh, the site of a pre-historic crannog, and
further along on the north side the ancient seat of Redcastle, founded by William the Lyon. This seaboard
between Fortrose and Tarradale, at the head of the Beauly
Firth, is one of the most delightful parts of Scotland for
climate, soil and scenery.

On the west coast, beginning at the south, Loch Alsh
lies between the narrows of Kyle Rhea and Kyle Akin,
and runs inland for about 7 miles, when it splits into
Loch Duich, nearly 5 miles long, on the right, and
Loch Long, about the same length but much narrower,
on the left. At their junction are the village of Dornie
and the ruins of the castle of Ellandonan on an islet
accessible on foot at low tide. The scenery here is very

fine, the "Five Sisters of Kintail" standing conspicuously at Loch Duich head. On the north side of Loch Alsh, near its mouth, is the terminus of the Dingwall and Skye railway, with a ferry three furlongs wide to Kyle Akin in Skye. Rounding the point, we pass a number of crofting townships and come to the village of Plockton near the mouth of Loch Carron. The loch is 12 miles long and has the village of Jeantown or Lochcarron (in Gaelic Torr nan Clàr) on its north side. From the mouth of Loch Carron, Loch Kishorn about 4 miles long branches to the left. Leaving these lochs, we round Rudha nan Uamhag, the south point of Applecross, with the Crowlin Isles in the offing, and pass along a somewhat indented coast to Applecross Bay. In this pleasant and retired spot St Mael-rubha founded his monastery in 673 A.D. The most northerly point of Applecross is Sròn an Iaruinn (Iron Cape). Here we enter Loch Torridon, 8 miles long, and by a narrow passage into Upper Loch Torridon which runs eastwards for 5 miles. The village of Shieldaig stands in Loch Shieldaig (Norse *síld-vík*, herring bay), a southern extension of Outer Loch Torridon. The coast of Gairloch northwards is low, and shows frequent stretches of the old sea beach as a raised terrace called in Gaelic *faithir*. There are many townships round the loch itself, which is about 3 miles long, and round the coasts of the big oblong peninsula north of it. The north-west point of this peninsula is Rudha Réidh, at which the coast turns eastwards, till after 5 miles we enter Poolewe, a fine broad inlet over 7 miles long, with well-peopled shores and at its head the village of Poolewe. It contains an isle described by Dean Monro in 1549 as "Ellan Ew haffe myle in

Gairloch-head (c. 1840)

lenthe, full of woods guid for thieves to wait upon uther mens gaire." The gardens of Inverewe and Tournaig with their tall fuchsia hedges and varieties of trees and shrubs show how favourable the climate of this district is to growth. North of Loch Ewe is another peninsula, in the lee of which is Gruinard Bay with Isle Gruinard,

Loch Toll an Lochain

and just beyond this comes Little Loch Broom, 7 miles long, with a very fine view of the jagged Teallach Hills, where in a valley on the east lies Loch Toll an Lochain. From Little Loch Broom Loch Broom proper, about 14 miles long, is separated by a peninsula which rises to 2032 ft. in Beinn Ghobhlach (the forked peak). Finely

situated on the north side of the loch is Ullapool. At its
mouth Loch Broom opens out into a wide island-studded
bay, 14 miles wide between Rudha na Cloiche Uaine
(Greenstone Point) on the south and Rudha na Cóigich
(Coigach Point) on the north. This little archipelago, at
some far-distant time part of the mainland, contains over
20 islets, of which only two or three are inhabited. The
largest is Tanera, about 1½ miles long. The coast of
Coigach is well occupied by small townships. On its
north side Enard Bay resembles Gruinard Bay; both
contain the Norse word *fjörðr*, a firth, bay.

On the west side of Lewis, the chief indentations are
the rather stern Loch Resort, Camas Uige (Uig Bay),
with pleasant sandy shores, and, most important of all,
Loch Roag, with its archipelago of isles clustering round
Great Bernera (Björn's Isle). Loch Roag is a splendid
roadstead, and its shores teem with interest. At its mouth,
on the south is Gallon Head, 5 miles north-east of which
is the small but lofty and precipitous isle of Berisay
(precipice-isle), where Neil Macleod held out for about
three years before he was taken and executed in 1613.
Near its head is Callernish. There are a few townships
south of Gallon Head, but many around Loch Roag and
on the low coast which stretches north-east towards the
Butt. The eastern side between the Butt and Stornoway
is also low and studded with townships as far as Cellar
Head. South of it comes Tolsta Head, between which
and Timpan head lies Loch a Tuath or Broad Bay,
guarded by the Eye Peninsula (in Gaelic Uidh, from
Norse *eið*, isthmus), about 7 miles long, with Ceann na
Circe, anglicised Chicken Head, at its south-west end.

Passing Swordale, we now come to the fine and busy harbour of Stornoway, opening southwards. The coast is thereafter rock bound, opening out into Loch Luirbost and Erisort, with a number of islets in their common mouth, and many townships on their shores. Rounding Rudha

Swordale Beach

na Càbaig (Kebock Head), we come to Loch Sealg or Loch Shell, after which the coast runs south-west, rocky and barren, to Loch Seaforth, which runs so far inland, at first north-west then north-east, that its head is within two miles of Loch Erisort. In the large peninsula thus formed is the deer forest of Park.

## 9. Climate and Rainfall.

Of the conditions that go to determine the climate of a district or country the chief are (*a*) latitude, (*b*) winds, (*c*) nearness to the sea, (*d*) mountains. We are apt to think of latitude, or distance north or south of the equator, as the most important of these, and on the whole the further north we go, the colder it gets. But the Channel Isles are further north than Vladivostock, which is ice-bound every winter. Nain in Labrador has an average temperature of 3·8 below zero in its coldest month, and Aberdeen 37·2 F., but both are in the same latitude. The British Isles have in fact a much more favourable climate than other places in the same latitude and similarly situated. The reason for this lies in the direction of the winds. Our prevailing winter winds are from the south-west: they come from warmer latitudes, and further they come across the Atlantic—a great reservoir of warmth. The winter winds of Labrador, on the other hand, blow across the frozen expanses of the north. Again, our climate is equable. We do not experience the great differences between summer and winter temperatures that are felt, for instance, in the interior of Europe or Asia. At Irkutsk, in southern Siberia, the mean temperature in January is about 8° F. below zero, and in July about 65° F., a difference of 73°. Edinburgh, which is in the same latitude (56° N.), has a mean January temperature of 40° F., and in July—August a mean of about 58° F., a difference of only 18°. The reason of the equability of our climate is our nearness to the sea. Water stores up the heat of the sun's rays slowly, and loses it slowly. Land stores up heat rapidly, and loses it rapidly. Thus with us a great part of the heat of summer is stored up in the seas that surround our coasts, to be given out slowly in winter, while a place remote from the sea is very hot in summer, but has no reserves of heat to draw on in winter. Such is the difference between a "marine" climate and a "continental" climate. The same effect, on a small scale, is felt in Scotland : places well inland are colder in winter than places nearer the sea.

The waters on our western coasts are warmer than those on our eastern coasts, and this used to be attributed to the influence of the Gulf Stream. The Gulf Stream, however, practically disappears to the east of the Newfoundland Banks, and seems to have little, if any, direct influence on our climate. The warmer waters on our western shores are part of a surface drift to the north-east, caused by the prevailing south-westerly winds acting on the warm top layer of the ocean.

In winter, the west coast of Ross is warmer than the east, and the inland parts are the coldest. In other words, the isotherms (lines connecting points of the same mean temperature) run roughly north and south. The mean January temperature of Lewis is about 41° F., the same as that of Dorset. Along the west coast at a little distance inland it is 40° F., almost that of the Isle of Wight. Near the watershed the mean temperature is 39° F., corresponding to the Weald of Sussex. The high-lying parts east of the watershed have a mean temperature of 38° F., like Essex. The parts about the head of the Moray Firth have a mean temperature in January of about 39° F.

In summer, the isotherms for Ross tend to run from south-west to north-east, and the east coast is warmer

than the west. For July, the isotherms connect Lewis
and Duncansbay Head, with a mean of 55° F. For the
west coast, the 56° F. curve connects Wick, Gairloch,
Applecross, Broadford. The 57° F. curve passes through
the North Sutor, and runs west-south-west to near the
head of Loch Duich, then south. The Black Isle lies
between 57° and 58° F. The mean temperature of the
south of England for July is 62° F. The mean annual
temperatures for four stations are as follows:

|  | Max. | Min. | Absolute Max. | Absolute Min. |
|---|---|---|---|---|
| Stornoway | 52·3 | — | 73 | — |
| Fortrose | 53·1 | 41·5 | 75 | 25 |
| Glencarron | 51·7 | 39·8 | 76 | 17 |
| Strathpeffer | 52·9 | 39·1 | 77 | 19 |

Rainfall is largely influenced by the direction of the
prevailing winds, and as the winds with us blow mainly
from the south-west and west, and come off the warm
Atlantic saturated with moisture, the west coast is wetter
than the east. If a country so narrow as ours were level,
presenting no obstruction to the winds, the rainfall would
be distributed fairly evenly, only that the west part would
be slightly wetter than the east. As it is, we have a high
watershed running approximately north and south, which
obstructs the winds and clouds in their passage, and con-
sequently the wettest part of Ross-shire is not Lewis, but
the part of the mainland just west of the watershed. The
wettest part of all is about Glencarron. When there is
heavy rain on the west, it is much lighter or even dry on
the east coast; conversely, with an east wind there is often
heavy cold rain in the east, while the west is dry and

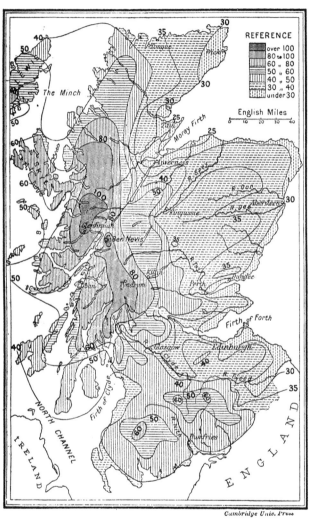

Rainfall Map of Scotland
(*By Andrew Watt, M.A.*)

sunny. The driest part of the county is a small strip running from east of Tain to west of Fortrose, with a mean annual rainfall of 25 inches. The line Ardgay—Alness—Beauly marks an increase to 30 inches. The 40-inch line runs through Garve, Cape Wrath, Butt of Lewis. At Loch Luichart there are 50 inches, and this isohyet (line of equal mean wetness) passes through the Eye Peninsula to Bragar. Achnasheen, on the watershed, has 60 inches, and between it and the line Kinlochewe—Shieldaig—Jeantown—Totaig there is an area with 80 inches. Most of Applecross, Gairloch, and Loch Broom runs between 50 and 60 inches. The warm west coast rain gets colder by the time it reaches the east ; rain from the east is cold from the start.

In Lewis there is little snow or frost. Snow seldom falls till after the beginning of January, and does not lie long ; so too on the west coast. In the central regions, and to a less extent in Easter Ross, snow often falls heavily and lies long. The great frost and snow of 1894–5 began on 28th December and lasted with little intermission till early in March. In several places the temperature fell to zero.

In the north, sunshine is not intercepted by smoke and artificial fogs, as is the case to a large extent in cities and areas of industry. The west is less sunny, i.e. more cloudy, than the east. The mean daily sunshine is 3·45 hours at Stornoway, and 3·8 hours at Fortrose. In July of 1920 the total at Fortrose was 253 hours, an exceptionally high record.

## 10. People—Race, Dialect, Population.

Ptolemy the geographer, who wrote about 140 A.D., names three tribes who appear to have occupied the mainland part of Ross-shire—the Carnonācae on the west, the Decantae eastwards from the Beauly river, and the Smertae in the north-east, in the basin of the Dornoch Firth. The name of this last tribe is still preserved in Carn Smeart, the Smertae's Cairn, a hill on the north side of Strathcarron in Kincardine parish. As those names are all Celtic, it may be inferred that before Ptolemy's time the Celts were the ruling people of these regions. They were not, however, the first inhabitants, for there is evidence of a still earlier race in Neolithic times, and between that Neolithic people and the Celts, there may have been still another race whose culture was of the Bronze Age. Many centuries after the Celtic conquest, the Norsemen began to settle in Lewis and later on the mainland.

The Neolithic people were dark, short, and of light make; the Celts were very tall, fair-haired and blue-eyed; the Norsemen were also fair and blue-eyed. The men of the Bronze Age were taller and stronger than the Neolithic people, and were most probably brown-haired. All these elements are represented in the present population. The dark Neolithic type is more common in Lewis than on the mainland. The fair-haired people of Ness in Lewis are supposed to be of Norse stock, but many of our fair-haired people are doubtless descendants of the old Celtic rulers. Brown hair, which is very common, may indicate mixture of races.

Of the language of the prehistoric people or peoples we know nothing for certain. The early Celts spoke British (represented now by Welsh), and traces of their language survive in the names of streams and places. Peffer, Carron, Conon are old British names, and the term Pit-, in Pit-kerrie, Pit-nellies, etc., was borrowed from the same source; it means primarily "a share." British was gradually displaced by the other Celtic language, Gaelic, introduced from Ireland by the Scots. The great majority of mainland names of places, and of very many Lewis names, are Gaelic. Norse names are not uncommon on the mainland; e.g. on the east side we have Dibidale, deep-dale; Amat, water-meet, confluence; Langwell, long field; Strath-rusdale where "strath" is Gaelic, and -rusdale is Hrút's dale; Cadboll, probably Cat-stead; Culbo, knob-stead; and most important of all, Dingwall, Thing-field, where the Norsemen held their "Thing" or court of justice. On the west are found names involving *vik*, a creek or bay, e.g. Dibaig, deep bay; Shieldaig, herring bay; Melvaig, bent-grass bay; Kirkaig, kirk-bay; Ard-heslaig, point of hazel-bay. Tanera, off Loch Broom, is haven-isle, Norse *hafnar-ey*, the *t* arising through Gaelic. Most of the principal names in Lewis are Norse, modified of course by transmission through Gaelic. Some of the leading Norse terms are (1) *á*, water, stream, as in Laxa, salmon river; Greeta, grit river; (2) *bólstaðr*, a home-stead, as Bosta, in Berneray; terminally -*bost*, as Garrabost, Shawbost, Melbost, Swanibost, Leurbost, Crossbost, Cala-bost; (3) *ey*, an isle: Pabay, Berneray, Taneray; (4) *gróf*, a pit, often in stream names: Allagro, Hundagro, Molagro; (5) *setr*, a residence, shieling: Shader, Grimshader, Hor-

Dingwall

shader, Linshader; (6) *nes*, a cape: Ness; Shilldinish, Aignish, Callanish; (7) *vatn*, a lake: Sandavat, Allavat, Ullavat, Tungavat; (8) *vágr*, a bay: Carloway, Stornoway, Leiravay, Loch Thamnabhaigh.

In 1921 the population of the county was 70,790 (33,668 males; 37,122 females). In 1911 it was 77,364. The population of the burghs is 11,426, and outside the burghs 59,364. In 1801 the total population was 56,318; thereafter it rose gradually to 82,707 in 1851, since when there has been a gradual decrease.

The number who speak Gaelic only is now 4860, or 6·87 per cent. of the total population. The number who speak both Gaelic and English is 35,810, or 50·59 per cent. of the whole. In 1911, Gaelic was spoken by 90 per cent. in Lewis; 85 per cent. on the western seaboard; 48·6 per cent. in Mid-Ross, and 29·8 per cent. in Easter Ross.

## 11. Agriculture.

Among Scottish counties Ross-shire stands third in total acreage, eleventh in extent of arable land. Of its total 1,977,248 acres, it has 111,213 arable, 26,581 under permanent grass, 955,117 of mountain and heath suitable for grazing, and 862,206 presumably unfit for grazing. Its chief products are wheat, barley, oats, potatoes, and hay, in all of which its average position as regards other counties is closely in proportion to its extent of arable land as compared with theirs. Easter and Mid-Ross and the Black Isle, with good soil, low average height above the sea, and a genial climate and exposure, are exceptionally

fine farming districts. When Thomas Pennant rode
through from Conon to the Dornoch Firth on 17th and
18th August, 1769, he found ripe apricots, peaches and
nectarines in the gardens of Brahan, and pronounced the
country east of Dingwall to be very well cultivated. The
valleys back of the hills were full of oats; the Laird of
Balnagown he declares to be the best farmer in the

Brahan Castle

country, as witness his wheat and turnips. Since that
time great progress has been made by means of enclosing
fields, reclaiming waste land, and improved methods and
implements, as well as by improved means of communication. The moister climate and thinner soil of the
western side make much of it better adapted for grazing
than for agriculture. How kindly the west is to vegeta-

tion appears from the success attained by Mr Osgood H. Mackenzie in his beautiful gardens and grounds of Inverewe and Tournaig. On 25th November, 1916, he notes " we can still produce specimens of most of the well-known flowers grown in British gardens—dahlias, begonias, gladioli, roses, and mignonette mixed up with some primroses, polyanthuses, and violets. How thankful I am that I live in a climate where the fuchsias are at their best in November."

The following figures give the acreage and crops for 1919 :

|  | Black Isle | Easter Ross | Mid Ross | South-Western and Western | Lewis |
|---|---|---|---|---|---|
|  | acres | acres | acres | acres | acres |
| Wheat ... | 56 | 1,127 | 188 | — | — |
| Barley ... | 2,754 | 1,752 | 1,692 | 21 | 2,259 |
| Oats ... | 8,238 | 11,384 | 10,492 | 1,997 | 3,486 |
| Potatoes ... | 790 | 1,380 | 788 | 745 | 3,542 |
| Turnips ... | 4,178 | 5,697 | 4,551 | 239 | 168 |
| Ryegrass ... | 11,078 | 15,558 | 13,126 | 2,165 | 325 |

The cattle on the eastern side are of various breeds, chiefly shorthorns, polled, and crosses. A considerable amount of fat beef is sent to the southern markets, and cattle from these districts take a high place at shows. On the west there is a considerable number of the shaggy, low, wide-horned West Highland breed. For 1919 the totals are : Eastern Districts, 24,647 ; Western Districts, 5475 ; Lewis, 11,661. The numbers for the west must have been greater formerly, for in 1793 there are stated to have been over 7000 black cattle in the parishes of

Applecross, Kintail, and Lochalsh alone. Kintail cattle were specially famed, hence the old saying "Ceann t-Sàil nam bodach 's nam bó." That was before "the big sheep" (na caoraich mhóra) were introduced.

The old native breed of sheep was small, and its mutton was very delicate. The larger species were introduced from the borders by that Laird of Balnagown who roused Pennant's admiration, much to the dissatisfaction and lasting loss of the people, who were so shamelessly cleared to make way for them. Sir George Mackenzie of Coul calmly remarks as a contemporary, "At the commencement of sheep-farming, the population was certainly very considerably reduced. There are some districts still too populous for improvement. The amelioration of the Highlands is only advancing." In 1792 the people attempted to drive the sheep out of the county altogether, but were stopped by military force. This year, which saw the triumph of the sheep, is known as "bliadhna nan caorach." Since then, the sheep have in their turn been largely evicted to make room for deer. In 1919 the number in the Eastern Districts was 136,375; in the Western Districts, 76,443; in Lewis, 53,998.

In the same year there were in the Eastern Districts 5879 horses, in the Western Districts 462, and in Lewis 1102. Pigs on the east numbered 2692; on the west 70; in Lewis 2. The loathing of pigs is general in the west, and is specially strong in Lewis. It is of very ancient standing, and may be safely considered pre-Celtic.

## 12. Manufactures, Mines and other Industries.

Apart from agriculture and fisheries, the industries of Ross-shire are not important. That iron was manufactured locally in early times appears from the numerous remains of primitive furnaces found all over the county, e.g. in Edderton, Rosskeen, Boath (Alness), at Loch a' Chroisg, in Glen Docharty, at Kinlochewe, Gairloch, and other places. The process doubtless went on continuously from a period before the Christian era. About the year 1608, Sir George Hay began the manufacture of iron at Loch Maree. At that time charcoal was used for smelting, and the particular attraction of Loch Maree was the great forests which grew around and near it. In 1610 Sir George Hay obtained a grant of the woods of Letterewe for his purposes. The ore was imported by sea from the south, but some local bog-iron ore seems to have been used. The works were at Letterewe, Talladale, and the Red Smiddy (a' Cheardach Ruadh) near Poolewe, and they produced wrought-iron, pig-iron, and articles of cast-iron, such as grates, and also cannon. A great business was done, up till about 1668, when cannon were still manufactured, and the reason of stopping it was most likely the exhaustion of the woods. It is reckoned that each furnace would use annually in the form of charcoal the product of 120 acres of wood. There were also extensive ironworks at Fasadh, on Loch Maree, where bog-iron ore was smelted with charcoal by lowland workmen; these were probably earlier than Sir George Hay's time.

Cottage tweed industries have been developed in Lewis, where it is reckoned that at least three-fourths of the crofters and cottars engage in the manufacture. In 1911

Handloom Weaving

Lewis produced about 300,000 yards of tweed, valued at £45,000. There are similar industries on a smaller scale along the western mainland. The manufacture of kelp, once a considerable industry in the northern Hebrides, has declined since the iodine, for which it is chiefly valuable nowadays, has been supplied from the nitrate fields of Chili. On the north side of the Cromarty Firth there are several whisky distilleries, and there is one near Tain. Local meal-mills have become few, even in Lewis, where there used to be a mill of simple construction on almost every stream. Local tradesmen, too, are fewer than they were, especially shoemakers, tailors, and weavers. Thus we are becoming more and more dependent on the south for the necessaries of life. There are tweed mills at Ullapool and Stornoway. A manure mill at Invergordon was closed in course of the war, and has not been re-opened.

The minerals known to exist in the county are iron, copper, asbestos, coal, and albertite, all in small quantities, and none of them worked at present. A vein of haematite occurs at Blackpark, on the Craigroy burn, Edderton, and close by are deposits of bog-iron ore and traces of an old ironwork. On the right bank of the Averon, about a mile above Ardross Castle, a large rock-face (Creag an Iaruinn) is exposed, which is rich in iron ore. South-west of it, on Loanroidge farm, are deposits of bog-iron ore and traces of an old furnace near them. Bog-iron ore rich in metal is found rather plentifully in Gairloch. There appear to have been early iron-mining operations in the burn of Allt na Mèinn (Altnamain), Edderton, near which is the so-called "Pictish Ironwork." Copper was mined at Kishorn in a pit which may still be seen, but as no traces of slag can be seen, it is probable that the ore was removed elsewhere for treatment. The work is not ancient, though there is a curious lack of information when and by whom it was carried on. Authorities also differ as to the value

of the ore. Asbestos is found in certain rocks near Dornie.
A very thin seam of coal crops out in Nigg. Albertite
occurs in the Strathpeffer district in veins that vary in
thickness from that of a sheet of notepaper to about three
inches. It looks somewhat like coal, and burns readily,
giving off clouds of smoke, with a bituminous odour, and
is considered to be an oxidation product of petroleum.

Sandstone is quarried extensively at Tarradale and
Lamington (near Tain), and also at Cullicudden, opposite
Invergordon.

## 13. Fisheries.

The total value of fish (excluding salmon) and shellfish
landed in Scotland in 1919 is estimated at £6,147,945, of
which Ross-shire contributed to the value of £523,907,
or over one-twelfth of the whole. The catch consists of
herring and whitefish, the former caught by nets, the
latter both by nets and by lines. The industry engages
about 3380 men, boys, and women; the value of the
vessels employed is about £234,791, and of their gear
about £158,050. Steam trawlers and drifters, motor boats,
sailing vessels, and small boats all play their part as means
of capturing the fish. Much of the catch is sold fresh, but
more is cured in various ways: herrings are salted or made
into kippers or bloaters; whitefish are dried, smoked, or
pickled.

The west coast is officially divided into three districts,
Stornoway, Loch Broom, and Loch Carron with Skye, and
of these Stornoway is the most important. There are two
herring seasons, the winter fishing from 1st January till

Stornoway Harbour

31st March, and the summer and autumn season from April or May onwards. During the winter season and the earlier part of the summer season, many stranger boats take part in the fishing; but by the end of June or sooner all these usually make for the east coast, leaving the west to the local fishermen. In 1919 more than half the Stornoway catch of herrings and about two-thirds of the

In the Fishertown, Cromarty

Loch Carron and Skye catch was made between January and March; in Loch Broom, however, December produced nearly two-thirds of the year's take. The white fishing of the west, which is also important, goes on all the year round, but here again in 1919 nearly half of the total catch was made in the winter period. In addition, the West Coast fishing (including Skye) in 1919 produced 241,640 lobsters, valued at £19,752.

The east coast is all included under the official district of Cromarty, which covers Cromarty itself, and the fishing villages of Inver, Portmahomack, Balintore, Shandwick, Avoch, and other smaller places. There are in all 45 district crews, and 381 men, boys, and women are employed in the industry. The fishing grounds are the Moray Firth and its branches, the Firths of Cromarty, Inverness, and Beauly. In the two latter sprats are fished. The herrings of these inner parts are small, and the Beauly Firth herrings are called (rather slightingly) "sgadain Poll an Ròid" by the people of the west. The official return of fish landed within the district in 1919 is 107 cwt. of herrings, valued at £53, but this represents only a small part of the catch, the bulk of which is landed at Inverness. The white fish landed in the district amounted to 13,327 cwts., valued at £24,534; shellfish, including 650 lobsters and 4500 crabs, were valued at £878.

There are many good salmon rivers on the mainland and in Lewis—the Grimersta river is specially noted— and salmon are caught by nets at a number of places along the coasts, but there is no available record of the number, weight, or value of the catch.

## 14. History.

The history of the county falls into four main periods: (1) the early period of Celtic supremacy; (2) the period of Norse influence; (3) the later period of Celtic institutions modified by feudalism; (4) the modern period, from 1746 onwards.

Cadboll Stone

5—2

(1) The early tribes had probably some political connection among themselves; at any rate when Columba visited Inverness about 565 A.D. the people were called collectively Picts, and the north as far as the Orkneys all obeyed one king, who ruled from Inverness. From about this time (565 A.D.) began the conversion of the North from paganism to Christianity through the work of clerics from Iona or from Ireland. Of Columba himself there is no trace on the mainland of Ross, but his contemporary Mo-Luag of Lismore is traditionally said to have founded a monastery at Rosemarkie, which became the centre of Christian influence in the eastern parts. There appears also to have been an early Christian settlement in the Tarbat district connected with a saint named Colman or Colmag. The number of old dedications and of beautifully carved Christian stone monuments in this whole region indicates the strength and culture of the Celtic Church here in these times. On the west, St Donnan of Eigg, slain by pirates in 617, is commemorated in Loch Broom (Kildonan) and in Kishorn (Seipeil Dhonnain), possibly too in Ellandonan. About 615 three Irish saints, Kentigerna, her son Fillan, and her brother Congan, came to the Lochalsh district, where there are dedications to all three—Cill Chaointeorn on Loch Duich, Cill Fhaolain in Kintail, Cill Chomhghain in Lochalsh. But the most important centre was Applecross, where a monastery was founded in 673 by Maol-rubha, who laboured in and from Applecross for forty-nine years, and clearly accomplished a great work. He died at Applecross in 722, and his grave there is still well known and reverenced. The monastery may have suffered from the Norsemen, but it was long a

Shandwick Stone

place of refuge or sanctuary, whence Applecross is called a' Chomraich. The other noted Comraich of the North was St Duthac's sanctuary or girth of Tain. The limits of both were marked by stone crosses. Duthac's name survives in Baile Dhubhthaich (Tain), Loch Duich on which is Clachan Dubhthaich (Kintail Church), and Cadha Dhubhthaich, a pass from Kintail into Glen Affric.

(2) The Norsemen appeared in the Hebrides at the end of the eighth century, but their settlement on the mainland took place later, about 875. They thoroughly colonised Caithness and part of Sutherland; Ross was not occupied so completely nor held so continuously. Norse power on the mainland was at its height under Thorfinn (d. about 1060), whose mother was daughter of Malcolm, king of Scots, and thereafter it declined, becoming restricted to Caithness. In Ross the native language was not displaced, though the people may well have been bi-lingual. As masters of the soil, the Norsemen doubtless made their subjects tend their flocks of sheep, and herds of cattle and horses, as well as till the land, while from those they did not enthral they would exact some form of tribute. Racially the Norse influence on the mainland of Ross is negligible.

(3) By the twelfth century Ross was again under native rulers, nominally subject to the king of Scotland, and in 1160 Malcolm Mac Eth (Mac Aoidh or Mackay, probably) appears as Earl of Ross. But there were risings against the authority of the king, and the region was not quiet till after Fearchar mac an t-Sagairt (Macintaggart, Priest's son) became earl about 1215. He was succeeded by four earls of his own line, when the earldom passed by marriage

to the Lesleys, father and son, and then, again by marriage, to the MacDonald Lords of the Isles. Donald, the founder of the new line, had to fight for his rights by marriage, whence the famous battle of Harlaw (Cath Gairbheach in Gaelic), in which, as a Scots noble wittily said, "MacDonald had the victory, but the government had the printer." The great power of the Lords of the Isles and Earls of Ross was dangerous to the Scottish Crown, and the imprudence of John, grandson of Donald of Harlaw, led to his final confiscation and forfeiture of lands and titles in 1494.

During the time of the earls the power of the Crown was exerted in the North on two notable occasions. In 1331 Thomas Randolph, Earl of Moray, holding a court of justice at Inverness, sent an officer to Ellandonan, where there were many misdoers. Fifty were seized, and in the words of Wyntoun :

> The hevyddis[1] off thame all
> Ware set wp apon the wall
> Hey on heycht on Elandonan,
> Agayne the come off the Wardan.

In 1427 King James I invited the Gaelic chiefs to a parliament or conference at Inverness, and there caused them to be arrested. Among them were Alexander, Earl of Ross and Lord of the Isles, and his mother; Kenneth More, leader of 2000 men (supposed to be chief of the Mackenzies); and Mackmacken, leader of 2000 men (perhaps chief of the Mathesons). These and others were imprisoned; several were executed. The Earl was soon

[1] heads.

released, and retaliated by burning Inverness. An army was sent against him, and after surrender he was warded in Tantallon. The king's treacherous *coup d'état*, far from pacifying the Highlands, led to the battle of Inverlochy (1431), in which his army was heavily defeated by Donald Balloch (freckled), a kinsman of the Earl and Lord of Dun-Naomhaig in Islay.

We hear of few internal conflicts in Ross itself, for the earls appear to have kept order within their bounds. In 1452, however, a fierce fight took place at Bealach nam Bròg, a pass on the western skirts of Wyvis, in consequence of an attempt by the Mackenzies to abduct a kinsman of the Earl, Ross of Balnagown. The Mackenzies suffered severe loss; on the other side the Dingwalls of Kildun lost, it is said, 140 men, and many of the Munros fell. In 1487 a quarrel between Mackenzie's heir apparent and Alexander MacDonald of Lochalsh, nephew of the Earl, led to the terrible battle of Park, near Strathpeffer, in which the MacDonalds were routed by the Mackenzies with great slaughter. It is said, probably with truth, that before the battle the MacDonalds burnt the church of Contin, with many women and children who had taken refuge in it. After the downfall of the Earl of Ross, Sir William Munro of Fowlis was chamberlain over his forfeited estates in Ross, and roused the resentment of Mackenzie, then living at Kinellan, by interfering with Mackenzie's property there in his absence. Mackenzie dared him to repeat the act when he was at home, whereupon Munro gathered 900 men, went to Kinellan, did there as he thought fit, and turned homewards without meeting opposition. Meanwhile Mackenzie had hastily collected about 140 men, whom he set in ambush on Druim a' Chait (the Cat's Back, near Strathpeffer), and fell on the Munros as they were passing through Bealach nan Corr (the Cranes' Gap), on the south side of Druim a' Chait. The Munros were defeated with heavy loss, and the courage and strategy displayed by the Mackenzies and their leaders won them additional prestige.

The Mackenzie fortunes advanced in the sixteenth and seventeenth centuries till the house of Kintail and its numerous branches ruled Lewis, nearly all the mainland west of Dingwall, most of the Black Isle, and even part of Easter Ross. The Munros held their ancient patrimony of Ferindonald, and acquired some land besides in Easter Ross. The Rosses remained within their old bounds in Easter Ross, including Edderton, the Carron basin, and Kyleside.

In the perplexing and trying period between the National Covenant of 1638 and the Restoration of 1660, the Earls of Seaforth and the Mackenzies were, as a rule, on the side of the king; the Munros were Covenanters, and the Rosses were on the same side, though perhaps with less ardour. Thus the Mackenzies fought at Balvenie (1649), and were disciplined for doing so by the Presbytery of Dingwall. The Munros and Rosses in 1650 helped to defeat Montrose at Carbisdale, now Culrain, on the Kyle of Sutherland. A considerable force of Mackenzies, Rosses, and followers of Sir Thomas Urquhart of Cromarty (but no Munros) took part in the disastrous battle of Worcester (3rd September, 1651). Many were slain, and many were sent across the Atlantic as slaves. From that time till 1660 Ross was under the military rule of Cromwell.

The bitter persecutions which followed the restoration of episcopacy under Charles II affected the county slightly as compared with other parts of Scotland. Five ministers refused to conform, and were ejected; two of these were repeatedly imprisoned on the Bass Rock and elsewhere. Conventicles were held in the East, one of which was

Urquhart Coat of Arms

broken up by soldiers; twelve laymen were fined in a total sum of £21,930 Scots.

In 1688 the West was Episcopalian and Jacobite; the East was Presbyterian and Williamite. No men from Ross fought at Killiecrankie, but Brahan and Castle Leod were garrisoned, and Seaforth was imprisoned till 1697. In the rising of 1715 Seaforth declared for James; the

Munros and Rosses were Hanoverian. Seaforth, who was at Sheriffmuir with his clan, fled to France, whence he returned in 1719 for the rising which was crushed at Glenshiel. His estates were forfeited, but chief and clan were pardoned in 1726. In 1745 some men fought for Prince Charles under the leadership of the Earl of Cromarty (George Mackenzie), but the county as a whole was either firmly for the government or lukewarm.

After Culloden certain measures were passed with a view to prevent further disturbances. The old obligation to military service (*feachd*) at the call of the superior was abolished: henceforward the tenure of land was to be by some form of rent. This was of course aimed at the power of the chiefs, and it was soon effectual in reducing the relation between superior and tenant to a purely commercial basis. Land held by charter tended to be regarded merely as a source of income; if the income could be increased by clearing off the tenants, there was nothing to prevent their being dispossessed. Heritable jurisdictions over criminal offences were abolished, and the holders of such received compensation. Henceforth all such offences were tried by courts of law—a great improvement on the old system. An Act was passed for disarming the Gaelic people and proscribing their national dress. This measure applied to all the clans, whether Jacobite or Hanoverian, and was naturally resented by all alike. As regards the tartan, Dr Samuel Johnson said justly that the enactment was "rather an ignorant wantonness of power, than the proceeding of a wise and beneficent legislation." The Act was rigidly enforced for ten years, and repealed only in 1782.

(4) Soon after 1760 began the period of "improvements," inaugurated in part by Admiral Sir John Lockhart Ross of Balnagown. Many of these were real and of lasting benefit, such as roads (made by government order); new crops, especially turnips; improved methods of farming; enclosing and reclaiming land; and planting trees. Others, in particular sheep farms and consolidation of small holdings into large farms, caused exceeding hardship at the time without yielding the permanent results anticipated. Sir John Lockhart Ross set the example of clearing off small tenants and cottars to make way for sheep, and the process went on till after the middle of the nineteenth century. Among the districts affected were, besides the uplands of the Balnagown property, Kildermorie in Alness parish, Kintail and Letterfearn, Culrain (1820), Strathconon (1840–1848), Glencalvie (1842–1846), Gruinards in Kincardine (1854), from which a host of young and old were thrust forth homeless to shift for themselves. The sheep farms, it is true, produced mutton and wool, but when these came to be imported from Australia the value fell, and sheep-walks were to a large extent converted into deer-forests. The policy of consolidation also caused many evictions, especially in the Black Isle, without much, if any, economic gain. It is to remedy the evils caused by the defective policy adopted after Culloden that recent legislation has been directed, ever since the Crofters' Act of 1886.

Lewis, or rather the whole "Long Island," is agreed to be the island called Dumna by Pliny and Ptolemy, a Celtic word connected with Gaelic *domhan*, deep. Memorials of its early inhabitants are found in the stones of Callernish, erected perhaps about 1500 B.C., and in the

brochs built by the Picts in (or before) the early centuries
of the Christian era. To what extent Lewis was influenced
by the Gael before the Norse invasion is not known.
The Norse period in Lewis ran from about 800 to 1266,
and for much of that time Man and the Isles formed a

Broch of Carloway

separate kingdom, dependent more or less on Norway.
In 1266 Magnus, son of Hakon, king of Norway, ceded
Man and all the Western Isles to Scotland for a sum of
4000 merks and a yearly payment of 100 merks. Norse
influence survives in the place-names and personal names
(MacLeod, MacAskill, MacAulay, MacIver), in the

number of Norse words borrowed into Gaelic, and to some extent in the physical appearance of the people in certain districts. Under Scottish rule Lewis fell to the Earls of Ross and later to the Lords of the Isles. Its native rulers were a branch of the great family of MacLeod, who held it till their own internal dissensions ruined them about 1600. From 1598 to 1610 is the period of the Fife Adventurers, a Company of lowlanders to whom James VI granted the island to exploit and colonise as was being done in Ulster. After three attempts at settlement, the native opposition proved too strong for the syndicate, and the survivors were glad to escape. Thereafter Lewis passed to the Chief of the Mackenzies, who received the title of Earl of Seaforth. The years 1652 to 1654 were stirring times in Lewis, and Stornoway Castle was held by the Commonwealth soldiers. The last Lord Seaforth died in 1815. In 1844 Lewis was sold to Sir James Matheson. Recently the whole island was acquired by a syndicate styled "the Lewis and Harris Welfare and Development Company," and organized by Lord Leverhulme.

In autumn of 1923 Lord Leverhulme made an offer of the whole of Lewis to the people as a free gift, in two parts—(1) Stornoway and the adjacent district within a radius of seven miles, (2) the rest of Lewis. The first part of this generous proposal was accepted ; the second part, after much consideration, was declined, owing to the fact that the expenses of administration and other services exceed the income derived from the land. As a consequence this second part of Lewis was offered for sale.

## 15.  Antiquities.

Prehistoric artifacts are conveniently classified as belonging to the Stone Age, the Bronze Age, and the Iron
Age. It is well to remember, however, that these "Ages"
are not sharply separated from each other, but overlap.
Stone implements continued in use long after the introduc-
tion of bronze, and may be found in use even among people
who are familiar with iron. General use of metals depends
not only on the knowledge of them, but also, of course,
on whether the supply of metals is plentiful or scarce. It
is well to remember, too, that the duration of the "Ages"
may vary in different countries or districts. The Stone
Age is divided into two main periods, the Palaeolithic or
early Stone Age, when implements were not polished;
and the Neolithic or new Stone Age, when implements
were polished. Whether Scotland has any relics of the
earlier period is a question under discussion, but Neolithic
artifacts are plentiful. In Ross-shire polished stone axes
and hammers have been found both on the mainland (in
Fearn, Alness, Gairloch, etc.), and in Lewis, as at Bragar
and Corrishader. Arrowheads, knives, and other implements of flint are common, particularly in the Black Isle—
evidently a favourite haunt of early man. Three very fine
arrowheads were found in a burial cairn at Stittenham.
Of several stone cups, one came from the great stone
monument of Callernish. A polished bracer of felstone
was found at Fyrish, and a round ornamented stone ball,
the size of a small orange, at Contullich.

Moulds for casting bronze axes and spearheads have
been found in Rosskeen and near the head of Strathconon,

a bronze spearhead at Londubh (Gairloch), an axe at Slattadale (Gairloch), an armlet at Little Loch Broom, a penannular ring with cup-shaped ends at Poolewe, and a number of objects among the Fendom sands. Bronze pins and a knife have been found in Edderton and at Knup in Lewis, and an important collection of bronze objects in a hoard at Adabrock, Lewis. Many finds, both of bronze and of stone, have disappeared without being recorded.

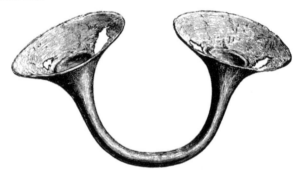

Penannular Ornament

Objects of jet (necklaces and rings) have been found in Lewis and on the mainland. Blue glass beads with yellow enamel spirals from Edderton are of Celtic manufacture. Fragments of Samian ware from a kitchen midden near Berie, Lewis; and two Roman copper coins (Tiberius and Nero) from Fortrose indicate intercourse of some kind with Roman Britain.

The county is rich in pre-historic structures. Burial cairns are more numerous on the eastern side than on the western. A large and massive long cairn stands towards the eastern end of Kinrive Hill; several chambered cairns further west have been wholly or partly destroyed. Scotsburn wood has several round cairns, and there are remains of two large chambered cairns in Boath. Others are in

Eagle Stone, Strathpeffer

Edderton, on Cnoc Navie (Neimhidh) in Rosskeen, and in the Black Isle. Na Clachan Gòrach (the Foolish Stones), on the heights of Dochcarty, are the nucleus of a chambered cairn. Cairns have been demolished at Stittenham, Cnoc Duchary (Alness), and elsewhere.

Stone circles, or remains of them, exist in Clare (Kiltearn), near Muir of Ord, and at Bealach nan Corr, near Jamestown. Some standing stones may be mentioned,

Callernish

such as the two large stones near Fodderty Church, Clach a' Mhéirlich (the thief's stone) west of Invergordon, and Clach an Tiompain at Strathpeffer, bearing the boldly cut figure of an eagle with certain symbols above it. In Lewis, the most remarkable arrangement of standing stones is that of Callernish, one of the most striking monuments in the British Isles. It consists of (1) a ring or oval of thirteen tall stones, with a still taller stone, 17 feet high, in the centre; (2) two lines of standing stones extending in a direction 10° east of north for 294 feet from the central pillar, forming an avenue about 27 feet wide; 19 of these stones remain in place, their average height being 10 to 12 feet; (3) three lines of standing stones extending southward, eastward, and westward from the oval for a distance from the central pillar of 114, 73, and 57 feet respectively; (4) the base of a burial cairn between the central pillar and the east or north-east side of the oval, containing a three-chambered grave with passage entering between two pillar-stones on the east side, the whole forming a cruciform stone-built chamber in the heart of the cairn, with the central pillar as a head-stone. The total number of stones remaining is 45 or 46; if all gaps were filled, the number would be about 60. East of the Callernish stones, two smaller groups stand on two knolls. Two similar groups are found near the head of Loch Roag. The Callernish stones and other smaller groups are called collectively "na Tursachan," a Gaelic plural of O. Norse *þurs*, a giant. We also find "Cnoc an Turs," the Giant's Hill, at Callernish, and "Clach an Turs," the Giant's Stone, at Point and elsewhere[1]. Martin says, "some of the ignorant Vulgar say, they were Men by Inchantment turn'd into Stones; and others say, they are Monuments

[1] Mr Kenneth Mackenzie, late schoolmaster of Shader, Barvas.

of Persons of Note kill'd in Battle." According to Geoffrey of Monmouth and others, Stonehenge was called "The Giants' Dance[1]"; its name in Welsh is "Côr y Cewri," the Giants' Circle. Single stone pillars are found near the head of Loch Carloway (3), on Bernera in Loch Roag (2), and in the Eye peninsula. More notable still is Clach an Truiseil at Shader, Barvas, $20\frac{1}{2}$ feet high and $15\frac{1}{2}$ feet round the base. Mr Kenneth Mackenzie explained "Truiseal" as O. Norse *þursholl*, Giant's Hill, the stone being named (now) after the hill, and this is doubtless correct.

Structures intended for habitation or defence or both are represented by the brochs, forts of various types, lake-dwellings, or crannogs, bee-hive huts, and hut-circles. Most of these belong to the Celtic period. The broch, called in Gaelic *dùn* or *caisteal* and in Norse *borgr*, was a round tower from 45 to 50 feet high, with a wall about 16 feet thick at the base, pierced by a narrow tunnel-like entrance with door half way in, and guard-chamber adjacent in the thickness of the wall. The wall itself at a height of about 8 feet becomes double, and the narrow space thus formed between the walls is divided into a number of floors or galleries one above another, separated by flooring of stone flags, to which access is given by a staircase which starts from the circular central court. This court is entered by the passage already mentioned, and is about 30 feet internal diameter, with two or three chambers off it, built in the thickness of the wall. The top is, now at least, open to the sky. The masonry is of

---

[1] In the original, "Chorea Gigantum"; he is translating the name from the Welsh of his day.

dry stone, but very strong. The brochs are known to have been occupied when the Romans were in Britain, and their beginnings may have been still earlier. Their occupants were well advanced in civilisation, and were agriculturists, seamen, and probably, many of them, pirates. On the mainland, there are remains of brochs at Birchfield and Kilmachalmag, on the Kyle of Sutherland; at Edderton; in Strathcarron (Kincardine); near Totaig on Lochalsh, and (probably) near the manse of Kintail (Dùnan Diarmaid); two on the south side of Loch Broom—Dùn na Lagaigh a ruin, and Dùn an Ruigh Ruaidh, a fine specimen with splendid masonry. Two brochs in Loch Carron (Dùn Carrannach and Dùn Ruigh-bhuadhchain or Revochan) seem to have disappeared, and in Boath and on the left bank of the Balnagown Water at Scotsburn are remains of circular structures which may have been brochs. In Lewis there are reported to have been 28, and their position is often indicated by Norse *borgr*, as Borronish (Fort-point) in Uig, with a broch called in Gaelic Dùn a' Chiuthaich; Tidaborra on Bernera Mór; Borve. The broch of Carloway is one of the finest specimens extant. Its wall is still about 30 feet high, and though only one side remains, the structure of the galleries is well preserved. Remains of several other brochs exist in the neighbourhood, one on a tiny islet : the broch men appreciated the advantages of Loch Roag.

Examples of hill-forts are found on Knockfarrel, Strathpeffer, a large oblong fortification which must have been once of first rate importance; Dunmore near Muir of Ord; and on the Ord of Kessock, a fort covering nearly three acres. All these are vitrified, that is, the stones of the

walls have been partly melted by fire so as to form a solid
mass like conglomerate rock, with glassy surfaces. On
Cnoc an Dùin, in Strathrory, there are remains of a large
fort or encampment called Dùn Gobhal, covering over
two acres. There is an interesting fortified headland at
Camas Mór, near the mouth of the Kanaird river in Loch

Crannog, Kinellan Loch

Broom, and several small forts in Lochalsh and near Dornie,
and in Gairloch. Of the Lewis forts, one of the most
notable is Dùn Othail on the east side south of the Butt.

Lake-dwellings or crannogs (from Gaelic *crann*, a tree)
have been in use from very early times till two or three
hundred years ago. One in Kinellan Loch, near Strath-

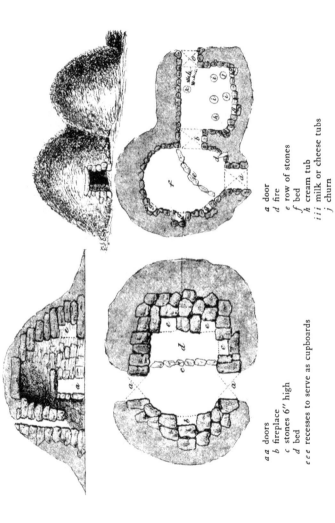

*a* door
*d* fire
*e* row of stones
*f* bed
*h* cream tub
*iii* milk or cheese tubs
*j* churn

**Both, Làrach Tigh Dhubhastail, Ceann Resort, Uig, Lewis**

*a a* doors
*b* fireplace
*c* stones 6″ high
*d* bed
*e e e* recesses to serve as cupboards

**Both, Cnoc Dubh, Ceann Thulabhig, Uig, Lewis**

peffer, was explored and fully reported on recently[1]. It appears to have been of mediaeval origin, and was used by the chiefs of the Mackenzies. Similar structures are found in Loch Beanncharan, Loch Tollie, and many other lochs.

Stone huts of the bee-hive type have been found near Tore in the Black Isle, one of which bore the interesting name Tigh a' Chruithnich, the house of the Cruithneach or "Pict." In Lewis a number of such structures exist entire on Morsgail moor and elsewhere. They are built exactly like the chambers of brochs, with courses of stones gradually overlapping inwards till the top is closed with one stone. In Lewis they are called in Gaelic *both*, while the modern shieling hut of stone and turf is called *bothan*.

Hut-circles are the round earth and stone foundations of ancient wattle huts, and usually stand on dry sunny slopes. The stone hearth or fireplace is readily discovered near the centre. They are common, and good examples are seen on the eastern slope of Kinrive Hill; near Arpafeelie in the Black Isle, where one of the group has right in the centre a stone with socket cut to receive the pole which supported the roof; and in Boath, where there was a regular village of them on the moor south of Ballone.

## 16. Architecture—(*a*) Ecclesiastical.

Of the wooden structures usual in the time of the early Celtic Church no trace remains. Nor have we in Ross-shire any certain specimens of the quaint dry-stone beehive

---

[1] Report by Mr Hugh A. Fraser, M.A., in *Proceedings of the Society of Antiquaries of Scotland*, 1916–17, pp. 48–98.

shaped cells which were sometimes used, probably by hermits; the huts of this type that are found on Morsgail moor in Lewis, were most likely shieling huts. Perhaps the oldest religious buildings we have are the tiny cells on the lonely isles of North Rona, Sùla-sgeir, and the largest of the Flannan Isles. They are oblongs, from about 7 to 14 feet long inside, built of dry stone, and so constructed that the flat roof is formed by laying heavy slabs of stone across. These cells, which were much reverenced, must have been the abode of recluses.

Next in antiquity to these are the small oblong chapels, built without cement, of Tigh a' Bheannaich (or, better, Tigh a' Bheannachaidh) near Gallon Head, and Dùn Othail, both about 18 feet long. Similar simple structures, but of later date and built with mortar, are Teampull na Crò Naomh at Galson, the church of St Aula at Gress (each about 19 feet long), and Teampull Pheadair at Shawbost (63 feet long). A stone above the door of St Aula's has the date 1685, in which year the church was probably repaired. St Columba's Church, on the Eye peninsula, near Stornoway, is a long narrow building of two compartments, divided by a thick wall. The eastern part is 62 by 17 feet, the western one 23 feet by 16 feet 3 inches. The character of the masonry and the style of the windows prove that the larger part is also the older. The masonry of the western division is pronounced to resemble Norman.

A more advanced type of church is that of St John the Baptist at Bragar, which was built with chancel and nave, though only 32 feet long inside. Its windows are flat-headed, and the broken chancel arch seems to have been

Tain from East

of the pointed form which succeeded the rounded Norman arch. This would indicate a date not earlier than the thirteenth century. St Columba's Chapel, on Eilean Choluim Chille in Loch Erisort, has a narrow lanciform window, pointing to a similar period. Teampull Mo-Luidh at Eoropie is a simple oblong, 44 by 17¾ feet, with a sacristy on north-east and chapel on south-east. The pointed arch of its east window implies that it is not earlier than the thirteenth century. Mr T. S. Muir remarks: "It is very entire, and judging from that, as also from its superior height and the somewhat refined nature of its plan, it is evidently much less antiquated than the other ecclesiastical remains in Lewis previously described."

Tain has three ancient buildings dedicated to St Duthac (Dubhthach), who died at Armagh in 1065 A.D. The first of these stands on a small knoll on the links, eastwards of the town, referred to in 1504 as "Sanct Duchois Chapell quhair he was borne." It is a simple oblong, 46 by 16½ feet internally, built of the granite boulders of the district. The south wall is ruinous; the other walls and the gables remain. The pointed window in the west gable suggests thirteenth-century work, but there are, it is said, indications that the original building may have been as old as the eleventh century. It is not unlikely that the chapel, as it stands, was constructed or re-constructed about 1253 A.D., when the relics of St Duthac were translated to Tain. It was burnt by Thomas MacNeil of Creich about the year 1427, and does not appear to have been rebuilt. The second chapel is an oblong structure of 32 feet by 13 feet inside the walls, which are now about 6 feet high. It had a lancet window and other features which point to an early date, and it is, apparently, at any rate, earlier than 1427. In 1504 King James IV made an offering "in Sanct Duchois Chapell in the Kirk-yarde of Tayne." Near this chapel stands the old Church of Tain, built by William, Earl of Ross, who died in 1371. In 1487 it was made a collegiate church, "ad instantiam Jacobi III Regis, in honorem Sancti Duthaci Pontificis." It is about 70 feet long by 22½ feet wide internally, and consists of a nave and choir, but without aisles. The east and west windows are large and filled with tracery, which in the case of the former was renewed in 1887, when the church was restored. The side windows looking south are also large and filled with tracery ; those in the north wall, exposed to the storms, are small plain lancets. The church is considered to be an example of middle pointed architecture, though later than work of that period in England. Here also James IV made an offering in 1504, on one of his many pilgrimages to St Duthac's shrine. His last visit was in 1513, before Flodden. In 1527 King James V made a pilgrimage to St Duthac's barefoot. The King's Causeway (Cabhsair an Rìgh), the name of the old road to Tain, remains as a memento of the royal pilgrimages.

Here we may mention some of the other old chapels or churches on the mainland of which remains exist. Cill Mo-Chalmaig (St Colman's Church) stands on the flat east of Kilmachalmag Burn, with an old burial ground quite neglected. Cill Mhoire (St Mary's Church), at the head of Loch Moire (Alness), is 42 feet by 18 feet inside, with walls about three feet thick cemented by very strong lime and now seven feet high all round. The door, 3½ feet

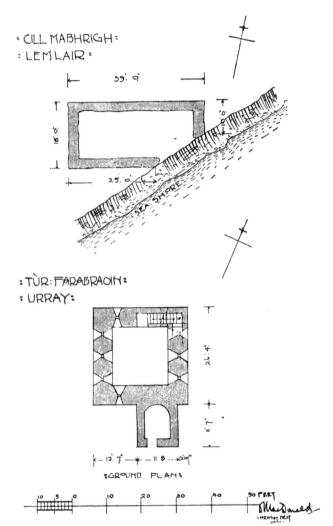

: CILL MABHRIGH :
: LEMLAIR :

39' 9"

18' 0"

25' 0"

SEA SHORE

: TÙR : FARABRAOIN :
: URRAY :

26' 4"

11' 7"

k— 12' 7" —*— 11' 9 —*2'10"

: GROUND PLAN :

10  5  0    10    20    30    40    50 FEET

MacDonald
: MENSOR DELT

Ground Plan of Lemlair Church and of Fairburn Tower

wide, and facing south is near the west end, and the only
window of which traces remain also faces south. It was
a place of great sanctity, and has a graveyard. At Cladh
Mo-Bhrìgh (St Bree's graveyard), on the sea shore about
a mile east of Dingwall, are the remains of a building
which was once the church of the old parish of Lemlair.
It stands about 13 feet above sea level, and so close to the
sea-shore that the south-east corner has disappeared owing
to coast erosion. The walls rise only slightly above ground
level, and are covered with turf. They measure 39 feet
by 18 feet externally. It would seem that these three old
churches (Kilmachalmag, Kildermorie, and Lemlair) are
typical of the older style of pre-Reformation church in
Ross. The earliest religious settlement at Rosemarkie was,
according to old tradition, by Mo-Luag of Lismore, who
founded a monastery there. In 710 A.D. Nectan, king of
the Picts, who desired to adopt the usages of the Church
of Rome, sent to Ceolfrid, abbot of Jarrow, asking to have
master-builders sent him to build a church of stone "after
the Roman manner," to be dedicated to St Peter. Closely
associated with Nectan was an Irish cleric named Curitan,
mentioned as an important bishop in 697, and elsewhere
styled "bishop and abbot of Ros mac Bairend." As
Curitan is reported to have died at Rosmarkyn and to
have been buried there, and as there are a number of
dedications to him throughout this region, and apparently
nowhere else, it is fairly certain that by "Ros mac Bairend"
is meant Rosemarkie, and that Curitan was abbot of
Rosemarkie. Further, as the foundation of the church
of Rosemarkie is traditionally ascribed to king Nectan, it
is probable that this was the church built by the master-

builders sent by Ceolfrid[1]. If so, it was the first church of stone built in the North.

When the Celtic Church was superseded by the Roman Catholic Church in the twelfth century, Rosemarkie became the seat of the bishop, who may have had a cathedral there, but in the thirteenth century, probably about 1235, the bishop's seat was changed to Fortrose or Chanonry (A' Chananaich). Here was erected a great building, of which part still remains. When it was complete, the Cathedral consisted of choir and nave, with south aisle thereto, lady chapel at the eastern end, great tower at north-west corner, and chapter-house at the north-east end. Contrary to the usual custom, it had no north aisle. What remains now consists merely of the south aisle and the chapter-house, and of these the latter is considered to be the older. The chapter-house is a rectangular building, 45 by 12 feet internally, of two storeys (crypt and chapter-house proper) standing by itself in the open space near the aisle, and now used by the Town Council. Externally it is plain, but its internal details, such as its dog-tooth mouldings, indicate thirteenth-century work. The part which has disappeared was probably of the same period, i.e. first pointed. The style of the south aisle is middle pointed. " The material, red sandstone, gave depth and freedom to the chisel, and the whole church must have been an architectural gem of the very first description. The exquisite beauty of the mouldings, after so many

[1] The choice lies between Rosemarkie and Restennet in Forfarshire; the fact that Curitan was Nectan's adviser and the practical certainty that he was abbot and bishop of Rosemarkie are in favour of the northern site.

years' exposure to the air is wonderful." It is composed
of two parts, an eastern part 41½ by 21 feet, and a western
part 56½ by 14¾ feet, with a bell turret at the angle where
the two parts meet. Both divisions are elegantly vaulted,
with rib edges and intermediate ribs. The east end had a
large traceried window of five lights. The south wall was
also pierced with traceried windows, now mutilated. The

Fortrose Cathedral

windows of the western part are less elaborate. Within
the aisle is a range of canopied monuments, the most
easterly of which is that of Euphemia, Countess of Ross,
who died before 1398 and is buried below. From certain
armorial bearings, it is inferred that the western part of
the aisle was erected either by the Countess or her son
Alexander, Earl of Ross and Lord of the Isles, and com-
pleted before 1439. "The whole of the architecture of the

aisle," say Messrs MacGibbon and Ross, "is of unusually good design, and the building is altogether quite unique and full of beauty and interest."

Dilapidation of the Cathedral began soon after the Reformation. In 1572 Lord Ruthven was granted "the hail leid quharwith the cathedral kirk of Ross is thickit," i.e. the whole lead covering the roof. We have the contemporary evidence of the Minister of Wardlaw as to the part taken in the demolition by Cromwell's men, when they were building their citadel at Inverness. " Most of their best-hewn stone was taken from Chanonry, the great Cathedrill and Steeple, the Bishop's Castle, to the foundation, rased." The citadel was begun in 1652; it was demolished in 1661. "It was a sacrilegious structure," says our authority, " and could not stand."

An interesting pre-Reformation parish church, abandoned about 1750, stands at Marybank, Logie Easter, with a neglected burying-ground. Its walls, built of small stones and very hard lime, are almost as firm as concrete, but the south wall is broken down. The church formed a somewhat narrow parallelogram, with a gabled wing projecting from the middle of the north side. The pitch of the three gables is steep. The north gable is pierced by two small windows. The east window was broad in proportion to its height, with a low-browed arch, and was divided by two red sandstone mullions into three compartments. The post-Reformation churches are rather featureless, but the conical belfry of Kilmuir Easter parish church deserves mention. It bears the inscription "Beigit 1616." Many of the existing parish churches were built abqut 1800, all on somewhat similar lines, plain and strong.

The churches built for the Free Church after the Disruption in 1843 were nearly all quite plain. Some of the later buildings are more ambitious in style.

## 17. Architecture—(b) Castellated and Domestic.

The word "castle" is ultimately from the Latin "castellum," a fortress; the early mediaeval castles were in fact strongholds, designed primarily for security against attack, and their essential feature was a strong wall enclosing an area of sufficient size. The site was often chosen with a view to natural strength and strategic position. The great wall was strengthened by towers; within it were the strong massive keep and other structures. The earliest castles in Ross were those of "Dunscath" and "Etherdover," erected by William the Lyon in 1179, for repressing insurrection. The former stood on the North Sutor, on the farm called after it Castle Craig; the name Dunskaith (Dun Sgath) is still extant, and traces of the fortification are said to be visible. "Etherdover," which means "between streamlets," was on the Beauly Firth, and occupied the site of the present mansion of Redcastle. Of the nature of these ancient fortresses nothing is known for certain, but it is not unlikely that they were earthworks, defended by palisades of wood.

The castle of Dingwall was a seat of the Earls of Ross. When and by whom it was built is uncertain, but in 1291 it was in the hands of the English, and its keeper was Sir William de Braytoft. Situated near the Peffery, on or near the spot occupied by the present Dingwall

Castle, it appears to have owed much of its strength to moats or ditches, readily constructed in a position low lying and naturally marshy. After the fall of the Earls of Ross, its keepers were appointed by the Crown till 1584, when it was granted to Sir Andrew Keith, who received the title of Lord Dingwall. Subsequently the custody of the castle appears to have been in the hands of the successive proprietors of the lands of Kinnairdie. The office of Constable of the castle is mentioned in 1700, and part of the building remained in 1792, but long before that date the old fortress had become a ruin.

The castle of Cromarty also was held by the English in 1291, its keeper being Sir Thomas de Braytoft. Later it became the seat of the Urquharts, hereditary sheriffs of Cromarty. Hugh Miller describes the building as it appeared before it was demolished in 1772. "It was a massy time-worn building, rising in some places to the height of six storeys, battlemented at the top, and roofed with grey stone. One immense turret jutted out from the corner, which occupied the extreme point of the angle, looking down from an altitude of at least one hundred and sixty feet on the little stream." There were other turrets of smaller size. The castle was defended by a dry moat and a high wall indented with embrasures, and pierced by an arched gateway. Within was a small court, flagged with stone. From the level of the court, a flight of stone steps led to the vaults below, while another flight of greater breadth, bordered on both sides by a balustrade, ascended to the entrance. The windows, small and narrow, and barred with iron, were thinly sprinkled over the front. How far Hugh Miller's description might apply to the

castle as it was in the thirteenth century is hard to say;
probably there had been changes and additions in the
interval. The site of the castle is now occupied by
Cromarty House.

On a rocky height overlooking the sea south of Avoch,
stood the strong castle of Avoch, also called Ormond
Castle. In 1297 it was held by the famous Sir Andrew
de Moravia (Sir Andrew Moray), and while the castles
of Inverness, Nairn, Urquhart, Dingwall, and Cromarty
were held by the English, it formed the centre and rally-
ing point of the patriotic party in the North. Andrew
Moray was killed at Stirling Bridge in September of
1297. His son Andrew Moray, regent during part of the
minority of David Bruce, died at Avoch in 1338, as
recorded by Wyntoun:

> Oure the Mounth than passit he
> Till Avawch in his awyne cuntre,
> And thare than endyt he his dayis
> As before the Cronykill sayis.

At a later date, the castle was held, but only for a short
period, by Hugh, Earl of Ormond, one of the Douglas
family, beheaded in 1455. The stones of the building
have been removed, and it is very likely that they were
used in the construction of Cromwell's fort at Inverness.
From the remains of the works, there appears to have been
a large rectangular, or approximately oval, enclosure, about
160 feet long and about 80 feet broad, with towers at the
angles, and several ditches and outworks. Of great interest
is the well or cistern, cut out of the conglomerate rock,
about eight feet in diameter and about twenty feet deep.
It was long choked with stones, but was cleared out about

twenty years ago, when many finely-hewn stones were discovered, but nothing else.

To the same period, or near it, belongs the castle of Ellandonan, on an islet accessible at low tide, at the junction of Loch Duich and Loch Long. The earliest mention of Ellandonan is that by Wyntoun, already noted (p. 71). In the fifteenth and sixteenth centuries

Ellandonan Castle

Ellandonan was held by the Mackenzies of Kintail. In the beginning of the sixteenth century there was trouble in the Isles consequent on the forfeiture of the Lord of the Isles and Earl of Ross, and the Earl of Huntly undertook to seize and garrison the castles of Strome and Ellandonan, as being " rycht necessar for the danting of the Ilis." In 1539 Donald Gorm of Sleat, besieging Ellandonan, was mortally wounded by an arrow shot from the castle; his followers are said to have burned the castle in revenge. In the attempted rising of 1719, the castle was held by a small company of Spaniards, and after being battered by three English men-of-war, it was blown up by a Captain Herdman. The great wall, of which the outline is still, or was recently, traceable, was in the form of a quadrilateral. The keep, in the north-east angle of the enclosure, was a regular rectangle measuring 57 feet by 43, with walls about 10 feet thick. Fragments of its north and south walls, of considerable height, were standing recently. A remarkable feature was a heptagonal tower on the east side, of 20 feet internal diameter, and open to the sky: within this was the cistern of the castle.

Though the castle of Strome was, as we have seen, an important stronghold, few details of its history are known. In the latter half of the sixteenth century, it belonged to Alexander MacDonald of Lochalsh, and remained in the hands of the MacDonalds till it was blown up by Kenneth Mackenzie of Kintail in 1602. It stood on a high rock, and a considerable portion of the walls was standing in 1893.

All these castles (Dingwall, Cromarty, Avoch, Ellandonan, Strome) were of the earliest type of regular Scottish feudal strongholds, characteristic of a period when the central authority was not yet strong enough to prevent local warfare, and when the territorial lords were bound to be able to guard themselves and their districts against all comers.

It is possible that along with these should be classed the Castle of Stornoway, but its early history is matter of tradition only. It was a place of great strength, and the key of Lewis, situated on the shore near the present landing-stage. Its site is marked approximately by a flagstaff, and all traces of the old building have been removed to make room for the harbour. The castle stood repeated sieges, notably in 1506 when it was taken by the Earl of Huntly, and in 1554, when it defied the artillery of the Earl of Argyll. It was dismantled by Cromwell's soldiers, and remained in a ruinous state thereafter. The present castle was built by Sir James Matheson.

A number of castles in the eastern part of the county belong to the fifteenth or the sixteenth century. Some of these have been modernised; others are ruinous. Of the latter, perhaps one of the oldest is Castle Craig, perched on a perpendicular rock on the south side of the Cromarty Firth nearly opposite Foulis. It is said to have been erected by the Urquharts of Cromarty, and it was at one time a residence of the Bishop of Ross. The top of the cliff was fortified by a wall, provided with round towers, and crenellated for defence. From the corbelling associated with cable moulding and dog-tooth ornament found on a parapet, it has been ascribed to the beginning of the sixteenth century, but from the absence of a fireplace in each room and the antiquated appearance of the one remaining, the date may be earlier. Fairburn Tower, near Muir of Ord, a Mackenzie stronghold, is a simple oblong keep, carried to a great height. The door giving access to the tower is on the first floor, and was reached by a circular staircase, now gone. On this floor was the hall, 16 feet square, with several wall-chambers. The room below the hall is vaulted, and could be reached only from above. The tower has five storeys, with two angle turrets at the top, and each room except that on the ground floor had a fireplace. The date is probably about 1600. Lochslin Castle, in Fearn,

Fairburn Tower

is still a conspicuous feature, though very ruinous. It was built on the L-plan, consisting of two towers joined at the corners, with a stair in the junction, and had three turrets. It stood about 60 feet high. Of Cadboll Castle, also in

Fearn, there was little left in 1840 except some vaults. It is reputed to have had the strange property that no one ever did or could die in it, and when the inhabitants as they grew old came to long for death, they had to be carried out. Ballone Castle, in Tarbat, appears to be a building of the late sixteenth century, though in 1840 it was said to have been then uninhabited "by any respectable family for 200 years." It was a strong place, built on what is called the Z-plan, with towers at the corners.

Redcastle, on the north side of the Beauly Firth, occupies the site of the castle of Eddirdover. The present mansion is modern, but incorporates part of an older structure. It is on the elongated L-plan, with turret stair-case in the angle, and corner turrets. Balnagown Castle, in Kilmuir Easter, was the seat of the Rosses of Balna-gown from early times till their line passed. It is a fine pile, consisting of an old west tower with high pointed roof and turrets, with additions of various dates. An eastern tower was added during last century in harmony with the older buildings. Kilcoy Castle, in Killearnan, is of the same type as Ballone, and belongs to the seven-teenth century. After long neglect it has been restored, and forms a very handsome residence. The best-preserved specimen of the dwellings of our Highland territorial families is Castle Leod, built by Sir Roderick Mackenzie in 1616. It is a modification of the L-plan, consisting of a main block with wing projecting beyond its north side and turrets. Newer additions have added a storey to the original building, finished with ornamental dormers and pyramidal turrets. The entrance door, which is in the addition, was ornamented with an elaborate coat of arms, now illegible. Brahan Castle, near Dingwall, was built by Colin, first Earl of Seaforth, about the same time as Castle Leod. It is a handsome spacious mansion, referred to in Gaelic poetry as " Brathann bhaidealach," " battlemented Brahan." The number of buildings erected by

Kilcoy Castle

the Mackenzies is an index of their power in Ross after the fall of the Earls.

The fine battlemented house of Tulloch, near Dingwall, was the seat of the Baynes, now of the Davidsons of Tulloch. Balconie Castle, near Evanton, is of somewhat similar style; it occupies the site of a residence of the Earls of Ross. Foulis Castle is the ancient seat of the chiefs of the Clan Munro. The old building, tauntingly described in a well-known Gaelic poem as "caisteal biorach, nead na h-iolair," "castle gaunt-peaked, the eagle's nest," was probably a tall lean tower of the Fairburn type. It was burnt about 1750, and rebuilt by, it is thought, a Dutch architect. The main building has a tower with the date 1754, and is flanked by two wings, each of which bears the date 1792. A local saying asserts that the castle has a window for every day in the year; as a matter of fact, some 280 windows have been counted. Other houses of similar style and date, built when the country was settling down after the rising of 1745, are Allangrange and Poyntzfield. The present house of Gairloch was built in or about 1738, near the site of an older mansion surrounded by a moat, and hence called "an Tigh Dìge," "the Moat House." The new building was slated, and was therefore called "Tigh Dìge nan Gorm Leac," "the Moat House of Blue Slabs." The old house had been thatched with heather. Conon House, formerly styled Logie House, is understood to be very old in part, but was added to in 1759 by Alexander Mackenzie, eldest son of Sir Alexander, who built Flowerdale House. The addition consisted of a wing on each side of the old tower. Coul House, another Mackenzie seat, was built in 1821. Ardross Castle was begun by Sir Alexander Matheson about 1850. About 1870 much of this building was removed and rebuilt from designs by Dr Alexander Ross in the old Scottish baronial style, but the turrets and the tower partake of the early French character, with cor-

bels and crow-stepped gables.  Another imposing structure
of recent date is Rosehaugh House, near Avoch, built in
the Renaissance style.

The Town Hall of Dingwall is described by authorities
on architecture as a massive example of a tolbooth of the

Town Hall, Dingwall

seventeenth century.  The door, as in Fairburn Tower,
is on the first floor, and is approached by a flight of steps,
giving access to a lobby which has the burgh court-house
on the right and the council chamber on the left.  Opposite
the door is a small wheel stair, leading to a room in the

tower, said to have been used as a debtors' prison. The ground floor is vaulted (as in Fairburn Tower), and contains cellars with doors to the outside. The cellar under the central tower was entered by a grated iron door under the steps. Here prisoners were kept, and the public communicated with them through the iron grating within the memory of persons living in 1892. An upper part was added to the tower in the beginning of last century, and in recent years the tower has been reconstructed.

## 18. Communications—Past and Present.

Wheeled vehicles were used in Scotland in very early times. The Caledonians of the first century A.D. fought from chariots, and native-made wheels of the Roman period are extant, well made and fitted with iron tyres. Early roads, however, were probably rough tracks or trails, and such they continued to be in Ross till the latter half of the eighteenth century. People walked or rode; internal traffic was carried on horseback. Unfordable streams, however, were sometimes bridged. The "bryg of Alness" existed in 1439; in 1649 there were applied to its "upputting" 200 merks of the vacant stipend of Contin. At the same time the burgh of Dingwall asked for part of the stipend "for putting up of their brigg." Ferries, such as those of Ardersier, Kessock, Invergordon, Cromarty, and Portincoulter (now the Meikle Ferry), were much in use.

The military roads constructed under General Wade (1725–1735) did not extend northward beyond Inverness. About 1761, some sort of road was made under his

Invergordon Ferry

successor, Major Edward Caulfield, from Contin to Pool-ewe, but whether it ever carried wheeled traffic is doubtful. In 1775, Taylor and Skinner's map shows a road from Inverness by Beauly and Tain to Portincoulter (the Meikle Ferry), with a branch from Dingwall toward Strathpeffer, another from about Nigg Station to Cromarty Ferry, and a third to Tarbat. In the Black Isle, roads run from Kessock Ferry to Tarradale, Conon, and Fortrose, and again from Conon to Cromarty, with branches to Rose-markie and Invergordon Ferries. On the West, the only road shown is that going by Glen Shiel and the head of Loch Duich over Màm Ràtagan to Glenelg.

The modern period of roads really begins with the Act for making Roads and Bridges passed in 1803, under which much work was done between 1807 and 1821, directed by the great engineer Thomas Telford. The Garve—Ullapool road was constructed about 1812, and in that year the old Bonar Bridge was built. To this period also belongs the road between Garve and Strome Ferry; the lateral roads on the West are considerably later. In 1808, Donald Ross was the first carrier between Inverness and Tain. In 1809, a coach or "diligence" began to run between Inverness and Tain, and it was possible to travel from Edinburgh to Carrol (in Sutherland), 215 miles, in 47½ hours. In 1818, "it is possible to travel from Edinburgh to John O' Groat's House without crossing a ferry or fording a river, or using a drag chain on a descent." Tolls were exacted for the use of the roads till 1866; some of the old toll-houses may still be seen.

The period of coaches continued till 1862, when the Highland Railway reached Dingwall. It reached Invergordon in 1863, and was extended to Bonar Bridge in 1866. The Dingwall—Strome Ferry branch was constructed between 1865 and 1868, and has since been extended to Kyle of Lochalsh. A branch has also been constructed between Muir of Ord and Fortrose. Cromarty, Gairloch, and Ullapool are still without railway connection, the difficulty being one of expense, not of engineering.

Before railways, all goods were of course imported and exported by sea, as they are still where railways do not exist. In 1792 "the goods imported to Dingwall from London, Glasgow, Leith, and other manufacturing and trading towns are carried in the London and Leith smacks," arriving "every three weeks or month at most." In 1839, a steamer (later two steamers) began to ply between Leith, Inverness, and Invergordon, which continues to be the chief shipping centre of the East side of the county. Most of the West still depends on sea communication, which is unfortunately neither cheap nor always adequate.

In Lewis good roads radiate from Stornoway to all parts of the island, and motor services make up, to some extent, for the absence of railways. Stornoway connects by boat with the railway at Mallaig and Kyle of Lochalsh; boats also ply between Stornoway and Glasgow, calling at many places intermediate.

## 19. Administration and Divisions.

In early times, the western part of Ross was known as North Argyll, or in Latin "Ergadia Borealis," for of old Argyll (meaning "coast of the Gael") extended from the

Clyde to Loch Broom. Hence it was that when the Norsemen came, they called the Minch "Skotland-fjörðr," Firth of the land of the Scots, for Scots and Gael were the same people. The Earldom of Ross extended from sea to sea, and on the south-west it was bounded by the Beauly river, and thus included part of what is now Inverness-shire. The boundary was changed in 1661 to its present line, but the parish of Kilmorack, situated in Inverness-shire, is still reckoned as part of Ross for ecclesiastical purposes. Lewis went with Ross as part of the possessions of the Lord of the Isles.

The district between the river Averon (or, as it is commonly called now, the Alness river) and Allt na Làthaid, to the east of Dingwall, has been known from of old as Fearann Domhnaill (Ferindonald), or "Donald's Land." It comprises the parishes of Alness and Kiltearn, and is the home of the Clan Munro. The origin of the name Ferin-tosh has been already explained (p. 4). This district was expressly excluded from the sheriffdom of Ross in the Act of 1661. Another sub-division, probably very old, is Coigach (a' Chóigeach) "the Place of Fifths," that part of the parish of Loch Broom which lies beyond Ullapool. There are, of course, many other smaller districts which have distinctive names.

The division into parishes must have been made soon after the erection of the Bishopric of Ross in the twelfth century: it would have been one of the first and most important tasks of the Bishop. The arrangement made then has been since modified to some extent, but not greatly. At present the parishes are as follows: (1) Easter Ross includes Kincardine, Edderton, Tain, Tarbat, Fearn (disjoined from Tarbat in 1628), Nigg, Logie Easter, Kilmuir Easter, Rosskeen. (2) Mid-Ross includes Alness, Kiltearn (including the old parish of Lemlair), Dingwall, Fodderty (including the old parish of Kinnettes), Urray (including Kilchrist), Urquhart or Ferintosh (including the old parish of Logie-Bride or Logie Wester), Contin, Loch Broom. (3) In the Black Isle District are Cromarty, Rosemarkie, Avoch, Resolis (formed in 1662 by uniting Kirkmichael and Cullicudden), Knockbain (formed by uniting Kilmuir Wester and Suddy), Killearnan. (4) The South-western district comprises Glenshiel (once part of Kintail), Kintail, Lochalsh, Lochcarron. (5) In the Western district are Applecross and Gairloch. (6) Lewis is divided into four parishes, Stornoway, Barvas, Uig, Lochs. The six districts as given above are those now in use for administrative purposes. It will be noticed that Mid-Ross includes one parish (Urquhart), which is geographically situated in the Black Isle, and one (Loch Broom) which belongs geographically to the West.

The mainland part of the county, including the burghs, is represented by one member of Parliament. Lewis is now part of the constituency styled "the Western Isles," which returns one member.

The county has a Lord Lieutenant and a large number of justices of the peace. The County Council, which has 55 elected members, attends to the finances, roads and bridges, public health, and general administration. Parochial matters, especially the administration of the poor laws, are in the hands of the parish councils, which are empowered to levy a rate for parish purposes. The royal burghs of Tain, Dingwall, and Fortrose, and the towns of Stornoway and Invergordon, are largely administered by their town councils, which deal with the property of the burghs, make and administer the by-laws, and impose the burgh rates. The police are under the county administration, the burghs paying a contribution annually to the county for their services.

The law is administered by a sheriff-principal and two sheriff-substitutes, one at Dingwall, who holds his court weekly at Dingwall and Tain, and monthly at Cromarty; the other at Stornoway. The head-quarters of the county constabulary are at Dingwall, and there are besides two districts, Tain and Dingwall, under inspectors.

By the Education Act of 1918 school-boards were abolished, and one Education Authority appointed for the whole county including Lewis. The Authority consists of 37 elected members, and has an Administrative Officer or Director of Education, and two school medical officers, one for the mainland and one for Lewis. There are besides local committees of management for each parish. The rate for education is levied by the parish councils, on a requisition made to them by the Education Authority. The Education Authority makes provision for School and University bursaries, and now, with secondary or higher grade schools in Tain, Cromarty, Fortrose, Dingwall, Stornoway, Invergordon, Plockton, and Ullapool, education is becoming more and more accessible to all who can profit by it.

## 20. Roll of Honour.

The surnames peculiarly distinctive of Ross-shire are Mackenzie, Munro, Matheson, and Ross. Indeed it may be taken as a rule that bearers of these names are connected with the county either directly or by descent. Some mention has been made already of great men closely connected with it in early times, though not born in it— Mo-Luag of Lismore, Mael-rubha of Applecross, Curitan of Rosemarkie, Donnan of Eigg, Congan, Fillan, and Kentigerna of Lochalsh and Kintail, and Dubhthach of Tain. The founder of the line of the Earls of Ross was Fearchar mac an t-Sagairt (Farquhar son of the Priest), surnamed O'Beòlain, and according to tradition a native of the county. He did service to Alexander II by suppressing a rising in Moray in 1215, and again in 1225 by giving him timely aid in Galloway. About 1225 he founded the monastery of Fearn, which, as we have seen, was originally situated east of Ardgay. The first Earl was evidently able and powerful. His daughter Christina was wife of Olaf, king of Man. He died about 1251, and was succeeded by his son William, who obtained a grant of Skye and Lewis from Alexander III, and died in 1274. The next Earl was also named William. In the Wars of Independence he sided alternately with the English and the Scottish parties, spent seven years in the Tower of London (1296–1303), and in 1305 violated the sanctuary of St Duthac of Tain by giving up to the English the wife and daughter of Robert the Bruce, who had taken refuge there. In 1308 he became reconciled to Bruce. His son Walter was a scholar at Cambridge in 1306, became a dear friend of Edward Bruce, and fell at Bannockburn. Isabella, Walter's sister, was betrothed to Edward Bruce in 1317, but he was killed in 1318. The next Earl was Hugh, who led an attack at Halidon Hill in 1333, and was slain there. The English found on his body St Duthac's shirt, which he wore for protection, and duly returned the relic to Tain. The saint had not forgiven the insult offered to his shrine by Earl Hugh's father. His son William was the last Earl of the O'Beòlain line. In 1346 he joined King David II at Perth to invade England, and there slew, or caused to be slain, Ranald of the Isles, in consequence of which both he and the Isles-men turned back from the expedition. King David was defeated and imprisoned in the Tower of London, and thereafter relations between him and the Earl were bad. William died in 1371; his daughter Euphemia married Sir Walter Lesley, and so brought the earldom into the family of Lesley. After her husband's death, she married Alexander Stewart, Earl of Buchan, "the wolf of Badenoch," son of King Robert II, and died Abbess of Elcho about 1394. She is buried in Fortrose Cathedral. Her son Alexander died in 1402, and through her daughter Margaret, who married Donald, Lord of the Isles, the earldom passed to the MacDonald dynasty.

Contemporary with the last Earl William was Paul MacTìre, "Paul the Wolf," who "was a valiant man, and caused Cathnes to pay him black maill....He got nyn scoir of cowes yeirly out of Cathnes for black maill so long as he was able to travell." He built a house on the promontory of Dun Creich which he left unfinished, dying "for displeasyr of his lost sone" slain in Caithness.

Paul MacTìre was chief of the Rosses or Clann Ghille-Andrais; his genealogy is given from an old manuscript in Skene's *Celtic Scotland*, III, 484. The site of his moot-hill near Tain is still known.

The native ruling families—Mackenzies, Munros, and Rosses—produced many good men, who were able administrators in their time, but for detailed accounts of them we must refer to the clan histories.

Gaelic literature owes much to Duncan Macrae of Inverinate, chief of his name, and a man of varied graces and accomplishments. In 1688 he began to compile a collection of Gaelic poetry, which is extant and known as the Fernaig Manuscript. Some of it is by himself and his brother, who was minister of Kintail, but most of it is by earlier writers, some of whom would otherwise have been unknown to us. He preserves poems by his own maternal great grandfather, MacCulloch of Park, by Donnchadh MacRaoiridh, who may have been bard to Seaforth, and by two lairds of Achilty, Alexander Mackenzie and his son Murdoch (Murchadh Mór mac mhic Mhurchaidh), all Ross-shire men. John Mackay, am Pìobaire Dall (1666–1754) was born in Gairloch and trained in piping under the MacCrimmons in Skye. His best known poem is Cumha Choire an Easa (the Lament of the Corrie of the Waterfall). His grandson, William Ross (1762–1790), was born at Broadford, but all his associations are with Ross-shire. He was schoolmaster in Gairloch, and one of the most accomplished Gaelic poets. Donald MacDonald (1780–1832), a native of Strathconon, and therefore called "am Bard Conannach," composed, among other poems, a spirited defiance to Napoleon Bonaparte.

John Mackenzie (1806–1848), born in Gairloch, where a monument stands to his memory (1878), compiled the great collection of Gaelic poetry entitled *Sàr Obair nam Bàrd Gàidhealach*, or *Beauties of Gaelic Poetry*, published in 1841 and several times since. He also wrote *Eachdraidh a' Phrionnsa*, a history of the rising of 1745, and edited a selection of Jacobite songs.

Donald Munro (fl. 1550), styled "High Dean of the Isles," was probably a native of Ferindonald. He wrote *A Description of the Westerne Iles of Scotland called Hybrides. Compiled by Mr. Donald Monro Deane of the Iles. 1549*, a valuable account, written in quaint old Scots. He was parson of Kiltearn and later (1574) of Lemlair, now part of Kiltearn. Sir Thomas Urquhart of Cromartie (1611–1660) was one of the most remarkable men of his age, a great scholar, a master of languages, and a lover of all that is recondite. His sketch of a universal language is pronounced to exhibit rare ingenuity, learning, and literary acumen. He is best known by his translation of Rabelais (1653), "one of the most perfect transfusions of an author from one language to another that ever man accomplished." Sir Thomas was a royalist, and suffered for the cause ; his death is traditionally ascribed to an uncontrollable fit of laughter on hearing of the restoration of Charles II. Alexander Munro (d. 1715 ?) son of Hugh Munro of Fyrish, was Principal of Edinburgh University. George Mackenzie, M.D. (1669–1725), a grandson of George, second earl of Seaforth, educated at Aberdeen, Oxford, and Paris, produced in three volumes *Lives and Characters of the most Eminent Writers of the Scots Nation*, and prepared a genealogical history of the families of Seaforth and

name of Mackenzie. He died of overwork at Fortrose.
Sir George Steuart Mackenzie (1780–1848), son of Sir
Alexander Mackenzie of Coul, and distinguished as a
mineralogist and geologist, is remembered by his *General
View of the Agriculture of Ross and Cromarty* (1813), a
valuable and interesting work, written from the point of

Hugh Miller

view of the landed proprietor by a man who had full
opportunities of gaining information, at a critical time in
the history of the Highlands. Hugh Miller (1802–1856),
probably the ablest and most interesting man that Ross-
shire has produced, was born in Cromarty. The story of
his youth and early manhood is told in his delightful book

*My Schools and Schoolmasters*. In 1840 he became editor of *The Witness*, and thenceforward lived in Edinburgh, exercising great influence on the side of the Free Church at the time of the Disruption. His chief works are *Scenes and Legends of the North of Scotland* (1835); *The Old Red Sandstone* (1841); *First Impressions of England and its People* (1846); *Footprints of the Creator* (1847); *Testimony of the Rocks* (1857); *Cruise of the Betsy* (1858). A monument to his memory overlooks Cromarty. Hugh Miller was a great English writer, a distinguished geologist, and a pious and patriotic man. His knowledge of the North lent weight to his denunciation of the Highland Clearances which Sir George Mackenzie regarded so complacently. His contemporary Sir Roderick Impey Murchison (1792–1871) was born at Tarradale and came of an old Ross-shire family. In his youth he served in the Peninsula under Sir John Moore, but after his marriage in 1814 he took to science, and became known over Europe as a geologist. Unlike Hugh Miller, he was always in easy circumstances and latterly wealthy. Sir Roderick contributed half of the endowment of the Chair of Geology in Edinburgh University. In 1818 he sold Tarradale, and thereafter he had little connection with the North. Hugh Andrew Johnstone Munro (d. 1885) was an illegitimate son of Hugh A. J. Munro of Novar, a connoisseur and Bohemian who formed a noted collection of pictures in London, known as the Novar Collection. The son was educated at Elgin Academy, Shrewsbury, and Cambridge, and became one of the most brilliant Latin scholars of his day. He was Professor of Latin at Cambridge, and will always be famed as editor and translator of Lucretius.

Alexander Mackenzie (1838–1898) was notable as politician, editor, and clan historian. Born on a croft in Gairloch, he settled in Inverness in 1869, and became editor and publisher of the *Celtic Magazine* and of the *Scottish Highlander*. An untiring worker, he produced

Hugh Andrew Johnstone Munro

seven clan histories of great genealogical value, and exercised much influence in ameliorating the conditions of the crofters.

Sir Alexander Mackenzie (1755–1820) was born at Stornoway; his father emigrated to New York in 1774.

He entered the service of a fur company, and in 1783
was sent to the North-West territory, at the head of
Lake Athabasca. From this point, on 3rd June, 1789, he
started on an expedition of discovery along the course of
the river flowing from Great Slave Lake, called after him
the Mackenzie River, and reached the point where it
enters the Arctic Ocean, returning on 12th September.
On an expedition westwards, which lasted from 10th July
1792 till 23rd August 1793, he was the first white man
to cross the Rocky Mountains and reach the Pacific Coast.
He published a book on his explorations, and was knighted
in 1802. Settling in Ross-shire, he bought an estate at
Avoch, and died at Mullinearn near Pitlochry as he was
journeying to Edinburgh.

Lieutenant-Colonel Colin Mackenzie (1754–1821) was
born at Stornoway, went to Madras in 1782 as a cadet in
the Engineer Corps under the East India Company, and
had a career of high professional distinction. He surveyed
Mysore (40,000 square miles), a work of great difficulty,
and finally became Surveyor-General of India. All along
he worked hard at the early history and archaeology of
the Deccan, and collected material to illustrate them.
He himself did not live to finish the arrangement of his
collections, which were sold for £10,000. Most of his
Sanskrit, Arabic, Persian, Javanese, and Burmese books,
coins, images, etc., are now in the India Museum, South
Kensington; his Manuscripts and Inscriptions are in the
Library of the Presidency College, Madras.

Sir George Mackenzie of Rosehaugh (1636–1691),
eldest son of Simon Mackenzie of Lochslin, was king's
advocate during the persecution of the Covenanters, when

for his severity he became known as "Bloody Mackenzie."
He had a fierce temper and a cruel disposition, and was
not above straining the law to secure convictions. He
wrote a number of books, legal and historical, all long
dead.

George Mackenzie (1630–1714), Viscount Tarbat,
first Earl of Cromartie, had a political career "perhaps
more variable and inconsistent than that of any other
Scottish statesman of his time." Of him Bishop Burnet said
that he had great notions of virtue and religion, but they
were only notions. Notwithstanding he was popular, and
when after Killiecrankie (1689) he was employed by the
government to treat with the Highland Clans, he showed
prudence, and helped materially to secure a settlement.
He was made Earl of Cromartie in 1703, and was a strong
advocate of the Union.

David Urquhart (1805–1877) was born at Braelangwell,
and educated on the continent and at Oxford. He became
distinguished as a diplomatist, especially in connection
with the Levant, and was trusted by the Turks as no
other was. He wrote much on political subjects, and is
said to have "impressed men of all opinions and nationali-
ties by his earnestness of purpose and the width of his
interests." He died at Naples.

Ross-shire has produced some soldiers of note, most of
them from the Clan Munro. Robert Munro, "the Black
Baron" of Foulis, was Colonel of a regiment under
Gustavus Adolphus, and died in 1633 of a wound received
at the battle of Ulm. Robert Munro, of the same family,
fought on the continent and in Ireland. He spent five
years in the Tower, from which he was released through

Cromwell in 1654. He died in Co. Down about 1680, and is said to have shared with another the honour of furnishing Sir Walter Scott with a model for the picture of Sir Dugald Dalgetty in the *Legend of Montrose.* Sir George Munro (d. 1693) of Culrain and Newmore served in the wars of Gustavus Adolphus, and fought at Lutzen. Afterwards he commanded with some distinction in Ireland, on the royalist side. Sir Hector Munro of Novar (1726–1805) had a notable career as General in India, where he suppressed the mutiny at Patna and won the decisive battle of Buxar, captured Pondicherry (1778), helped to gain the victory of Porto Novo, and captured Negapatam. In India he amassed a large fortune. It was he who erected the monument which so conspicuously crowns the hill of Fyrish, in the form, it is said, of the gate of Negapatam, and also structures which stand on lower eminences west of Fyrish. Sir Hector Macdonald (1852–1903), a native of the Black Isle, enlisted as a private soldier, served in the Afghan war, and was promoted from the ranks. He fought in the Boer war (1881), and in the Sudan (1885–1898), and won great distinction at Omdurman. In the South African war he commanded the Highland Brigade. In 1901 he was sent to a command in India, and in 1902 was transferred to Ceylon. Macdonald had great qualities, and the regard in which he was held is shown by the national monument to his memory which stands at Dingwall.

Among notable clergymen born in the county may be named Eneas Sage, born in 1694 at Chapelton, Redcastle, and minister of Loch Carron from 1726 till his death in 1774, a man of uncommon strength of character and of body, who was able and willing to convince his erring parishioners by physical means when moral suasion failed. His son, Alexander Sage (1753–1824), born at Loch Carron, was minister of Kildonan when that strath and Strathnaver were cleared of the people to make way for sheep, and some of the dark story is related by his grandson Donald Sage, who at that time was minister at Achness

Rev. Gustavus Aird, D.D.

in Strathnaver. Donald Sage became minister of Resolis, and wrote an account of his grandfather, his father, and himself, with reference to their times, under the title *Memorabilia Domestica*, a delightful and very valuable book. Charles Calder Mackintosh, D.D. (1806–1868), was born at Tain, of which his father, Dr Angus Macintosh, was minister. He was a good scholar and much esteemed as a preacher and as a man. Gustavus Aird, D.D. (1813–

1898), born at Heathfield in Kilmuir Easter, was minister at Croick and thereafter at Creich, with which his name is always associated. He was a courtly gentleman of the old school, a highly original and arrestive preacher, with much quiet humour, and was greatly revered. James Calder Macphail, D.D. (1821–1908), born in Loch Broom, was

Rev. James Calder Macphail, D.D.

long minister of Pilrig Free Church, Edinburgh. For many years he conducted a scheme, originated by himself, for providing "Grammar School Bursaries for Gaelic speaking young men." Dr Macphail's scheme ceased only when it was no longer necessary; while it lasted it was of very great service.

## 21. The Chief Towns and Villages of Ross-shire.

(The figures in brackets after each name give the population in 1921, and those at the end of the sections give references to the text.)

**Alness**, *Alanais*, astride of the river Averon, 10 miles north of Dingwall. The greater part of the village is on the Rosskeen side, but the orginal " Alness " is about a mile west of the river, where the parish church is. Alness has a Town Hall, reading and recreation rooms, several hotels, and whisky is manufactured in the neighbourhood. (pp. 41, 113, 114.)

**Ardgay**, *Ard-gaoithe*, Wind-point, a small village on the Ross-shire side of Bonar Bridge, with a station called Bonar Bridge. It is pleasantly situated, and has a good hotel. Here is held the annual winter market known as the " Féill Éiteachain," traditionally said to follow a certain quartz block, now built into the wall of the hotel. (p. 39.)

**Avoch**, *Abhach*, usually pronounced Obhach, Stream-place, a thriving fishing village in the Black Isle, near Fortrose. (pp. 42, 66, 100.)

**Balintore**, *Baile an Todhair*, Bleaching Stead, a fishing village in the parish of Fearn, with a harbour. (pp. 40, 66.)

**Barbaraville**, *an Cladach*, the Shore; its eastern part is Port Fhlich, Wet port or ferry, on the Cromarty Firth 3½ miles north of Invergordon.

**Cromarty** (1126), *Cromba*, on the southern shore of the Cromarty Firth, near its mouth. It is a parliamentary burgh and a sea-port, strongly fortified during the War as commanding the entrance to the Firth. Here in 1802 was born the geologist and man of letters Hugh Miller, whose monument occupies a conspicuous position. His parents' house is still standing, and serves as a Hugh Miller Museum. (pp. 40, 66, 99, 109, 120.)

Sculptured Stone, Dingwall

**Dingwall** (2323), *Inbhir Pheofharain*, Peffer Mouth, royal burgh and capital of the county, at the head of the Cromarty Firth. Occupying an important strategical position, Dingwall has been a place of note since very early times. Here the Norsemen held their Thing or court of justice during their occupation, and in still earlier times Knockfarrel, close by, was a great native stronghold. It is now the business centre of Mid Ross, and has weekly auction sales of stock. It is also the centre of administration for the county, and has an excellent central school, and three newspapers. Opposite the academy there is a mound formerly surmounted by an obelisk about 56 feet high, erected by the first Earl of Cromarty (d. 1714), who is buried at its base. The earth forming the mound is traditionally said to have been brought in creels on women's backs from the various estates which formed the county of Cromarty. The obelisk leaned considerably from the perpendicular, and in 1916 it was taken down, and replaced by a rather stumpy pillar. (pp. 13, 40, 54, 94, 108, 114, 115.)

**Dornie**, *an Dornaigh*, the place of hand-stones, i.e. rounded pebbles, a village, partly fishing and partly crofting, at the junction of Loch Duich and Loch Long. (p. 42.)

**Evanton**, *am Baile Ur*, also, *am Baile Nodha*, both meaning New-town, in Kiltearn, close to the famous Allt Granda, and between that stream and the Skiach water. The poem *Cabar-féidh* boasts that Seaforth in 1718 burned "am Baile Nodha" when ravaging Ferindonald:

> Am Baile Nodha 'na shradagan,
> Is 'na lasair anns na speuran,
> An uair dh'eireadh do chabar ort. (pp. 13, 41, 106.)

**Fortrose** (963), *a' Chananaich*, the Chanonry, formerly the ecclesiastical centre of Ross, and the seat of the Bishop. A royal burgh, it is situated on a terrace overlooking the Moray Firth, and is a delightful residential place, much frequented in summer and autumn. Some of the Earls of Seaforth lived chiefly at Fortrose. (pp. 37, 42, 115.)

**Hilton**, *Baile a' Chnuic*, Hill-town, a small fishing village near Balintore. (p. 40.)

**Inver**, *an Inbhir*, a fishing village about half way between Tain and Portmahomack. (pp. 40, 66.)

**Invergordon** (1384), *an Rudha*, the Point; in full, *Rudha Aonach Breacaidh*, the Point of Breakie Fair, called Invergordon by a Gordon proprietor of the eighteenth century. It is a long wide street, with short side streets leading mostly to the harbour. During the War, so many men were engaged on national work that additional houses had to be built for them at the neighbouring village of Saltburn, but Invergordon has now been abandoned as a naval base and is as it was before, except that its manure works have not been restarted. A fair amount of trade is done through its harbour, and it has an auction mart. (pp. 40, 115.)

**Jamestown**, *Baile Sheumais*, a tiny crofting village near Strathpeffer, in which visitors to the Spa often take rooms.

**Jeantown**, *Baile Séine*; formerly *Torr nan Clàr*, Knoll of the Boards, a large crofting and fishing village on the north side of Loch Carron. (p. 43.)

**Kessock**, *Ceiseig*, on the north side of Kessock Ferry, a village of crofters, fishermen, and pilots. (pp. 42, 109.)

**Maryburgh**, *Baile Màiri*, near Dingwall; near it is the Seaforth Sanatorium.

**Milntown**, *Baile Mhuilinn*, an old-world village, with meal mill, village green and cross, and village well. The older people called it *Baile Mhuilinn Anndra*, "Andrew's Mill-town," from Black Andrew Munro, who built Milntown Castle about 1500 and has left some dark traditions. It depended first on the castle, now quite gone, and thereafter on New Tarbat House—a feudal village.

**Muir of Ord**, *am Blàr Dubh*, the Black Muir, where an important market used to be held, till the auction marts put it out of date. (pp. 85, 103.)

**Munlochy**, *Poll-lochaidh*, changed from *Bun-lochaidh*, Loch foot, a pleasant country place at the end of Munlochy Bay. (p. 42.)

**Plockton**, *am Ploc*, the Lump; *Ploc Loch-Aillse*, Lump of Lochalsh, a prosperous crofting and fishing village of Lochalsh, with bathing and boating. It is rather a favourite holiday resort. (pp. 43, 115.)

Strathpeffer with Ben Wyvis in background

**Poolewe,** commonly in Gaelic *Abhainn Iù*, in Gairloch ; a crofting village, with a small but good hotel. (pp. 17, 43.)

**Portmahomack,** *Port mo Cholmaig*, Saint Colman's Port, has some shipping and fishing, a golf-course, and a Carnegie Library. Three miles from it is Tarbat Ness Lighthouse. (pp. 40, 66.)

**Rosemarkie,** *Ros-maircnidh*, Point of Marknie, i.e. Horse-burn, the earliest seat of Christianity in Ross, and the seat of the first Bishops of Ross. Though eclipsed to some extent by Fortrose, it is a charming place, and has interesting relics, including the well-known Celtic sculptured stone. (pp. 42, 94.)

**Saltburn,** *Alltan an t-Salainn*, east of Invergordon, of which it is practically a suburb. The name dates from the time of the salt duties, when smuggled salt used to be landed and concealed about the banks of the streamlet. The place is one long row of houses facing the firth.

**Shieldaig,** *Sìldeig*, from Norse *síld-vik*, Herring Bay, a crofting and fishing village at the head of Loch Torridon.

**Stornoway** (4079), *Steòrnabhagh*, Norse, Steerage Bay, a royal burgh, the capital of Lewis and the chief seat of the fishing industry on the West Coast. It is a well-built modern town, and has an excellent school, the Nicolson Institute. Near it is a fine golf-course. (pp. 47, 62, 102, 115.)

**Strathpeffer,** *Strath-pheofhair*, Strath of the Peffer, a well-known health resort, supposed to alleviate or cure rheumatism by means of its waters. The "Wells" are now the property of a syndicate, which has done much to increase the fashionable character of the place. (p. 15.)

**Tain** (1551), *Baile Dhubhthaich*, Saint Duthac's Town, at one time the capital of Ross, now the business centre of the district east of Balnagown Bridge. Tain, like Fortrose, owed its origin to the Church. It grew around St Duthac's shrine, and benefited by the privileges attached to it, and by the pilgrims who frequented it. Though quiet, it is really a prosperous town, and its fine air and surroundings make it a desirable place to live in. It is well built, and has a very pleasing Town Hall. Tain Academy is a good Secondary School. The golf links are said to rival those of Dornoch. (pp. 32, 39, 40, 92.)

Ullapool

**Ullapool,** *Ulabol,* Norse, Ulli's Stead, on Loch Broom, was founded in 1788 by the British Fishery Society, and was intended to be a spacious and beautiful town on a regular plan. Though it has not fulfilled what was expected of it, its fine position, well-built houses, and well-arranged streets make it a pleasant little town, and it has good facilities for fishing, boating, and bathing. It has a tweed manufactory. (pp. 46, 62, 115.)

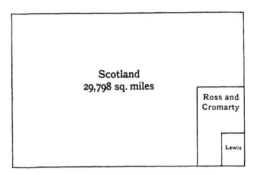

Fig. 1. Area of Ross and Cromarty (3089 square miles)
compared with that of Scotland

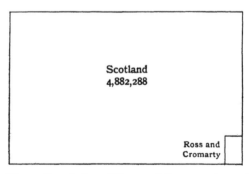

Fig. 2. Population of Ross and Cromarty (70,790)
compared with that of Scotland in 1921

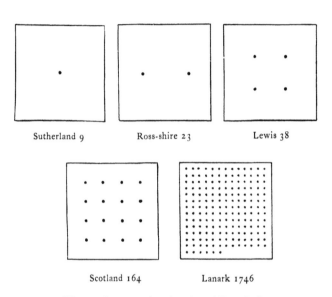

Sutherland 9     Ross-shire 23     Lewis 38

Scotland 164     Lanark 1746

Fig. 3. Comparative density of Population
to the square mile in 1921

(Each dot represents 10 persons)

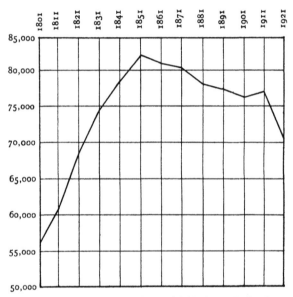

Fig. 4. Graph showing rise and fall of population in
Ross and Cromarty 1801—1921

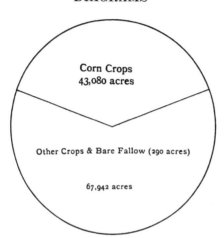

Fig. 5.  Area under Cereals compared with other crops
in Ross and Cromarty in 1921

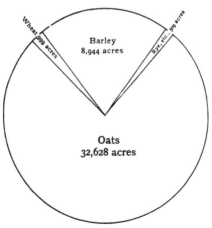

Fig. 6.  Proportionate areas of chief Cereals in Ross
and Cromarty in 1921

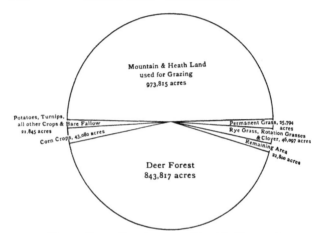

Mountain & Heath Land
used for Grazing
973,815 acres

Potatoes, Turnips,
all other Crops & Bare Fallow
21,845 acres

Corn Crops, 43,080 acres

Permanent Grass, 25,794 acres

Rye Grass, Rotation Grasses
& Clover, 46,097 acres

Remaining Area
32,800 acres

Deer Forest
843,817 acres

Fig. 7.  Proportionate areas of land in Ross and
Cromarty in 1921

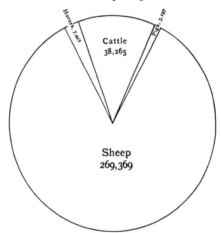

Horses, 7,442

Pigs, 3,197

Cattle
38,265

Sheep
269,369

Fig. 8.  Proportionate numbers of Live Stock in
Ross and Cromarty in 1921